AF744834

HYBRID ELECTRIC VEHICLES

Edited by **Teresa Donateo**

Hybrid Electric Vehicles
http://dx.doi.org/10.5772/66000
Edited by Teresa Donateo

Contributors

Wei Wu, Julius Partridge, Richard Bucknall, Muhammad Aziz, Nicolae Florin Jurca, Mircea Ruba, Enhua Wang, Fuyuan Yang, Minggao Ouyang, Aurelio Soma', Yannis L Karnavas, Ioannis Chasiotis

Published by InTech
Janeza Trdine 9, 51000 Rijeka, Croatia

© The Editor(s) and the Author(s) 2017
The moral rights of the editor(s) and the author(s) have been asserted.
All rights to the book as a whole are reserved by InTech. The book as a whole (compilation) cannot be reproduced, distributed or used for commercial or non-commercial purposes without InTech's written permission. Enquiries concerning the use of the book should be directed to InTech's rights and permissions department (permissions@intechopen.com).
Violations are liable to prosecution under the governing Copyright Law.

Individual chapters of this publication are distributed under the terms of the Creative Commons Attribution 3.0 Unported License which permits commercial use, distribution and reproduction of the individual chapters, provided the original author(s) and source publication are appropriately acknowledged. More details and guidelines concerning content reuse and adaptation can be found at http://www.intechopen.com/copyright-policy.html.

Notice
Statements and opinions expressed in the chapters are these of the individual contributors and not necessarily those of the editors or publisher. No responsibility is accepted for the accuracy of information contained in the published chapters. The publisher assumes no responsibility for any damage or injury to persons or property arising out of the use of any materials, instructions, methods or ideas contained in the book.

Publishing Process Manager Martina Usljebrka
Technical Editor SPi Global
Cover InTech Design team

First published June, 2017
Printed in Croatia
Legal deposit, Croatia: National and University Library in Zagreb

Additional hard copies can be obtained from orders@intechopen.com

Hybrid Electric Vehicles, Edited by Teresa Donateo
p. cm.
Print ISBN 978-953-51-3297-4
Online ISBN 978-953-51-3298-1

We are IntechOpen, the world's largest scientific publisher of Open Access books.

3,250+
Open access books available

106,000+
International authors and editors

112M+
Downloads

151
Countries delivered to

Our authors are among the
Top 1%
most cited scientists

12.2%
Contributors from top 500 universities

CLARIVATE ANALYTICS
BOOK CITATION INDEX
INDEXED

WEB OF SCIENCE™

Selection of our books indexed in the Book Citation Index
in Web of Science™ Core Collection (BKCI)

Interested in publishing with us?
Contact book.department@intechopen.com

Numbers displayed above are based on latest data collected.
For more information visit www.intechopen.com

For EU product safety concerns:
IN TECH d.o.o., Prolaz Marije Krucifikse Kozulić 3, 51000 Rijeka, Croatia,
info@intechopen.com or visit our website at intechopen.com.

Meet the editor

Teresa Donateo is an associate professor of Fluid Machinery, Energy Systems, and Power Generation at the University of Salento since 2014, teaching courses on Fluid Machinery and Hybrid-Electric Power Trains. She attended the School of Material Engineering at the University of Lecce (Italy) and graduated in 1999. In November 2001, she joined the Faculty of Engineering at the University of Salento as an assistant professor and in July 2003 received her PhD degree from ISUFI in the field of combustion and energy conversion. She has been collaborating since 2005 with the Ohio State University, Columbus, OH, and since 2001 with major automotive and aircraft industrial partners. Her research topics are simulation, design and optimization of internal combustion engines, and hybrid-electric power trains.

Contents

Preface

Hybrid electric vehicles are a mature technology for road applications and of increasing interest in other fields including aircraft. They rely on the use of two different forms of energy storage (a fuel and a battery) and try to optimize the energy flows between the corresponding energy converters (electric machines and thermal engines). Several studies in literature have pointed out the advantages of HEVs versus conventional vehicles in terms of fuel economy and environmental impact. They are mainly due to the flexibility in the choice of engine operating point that allows the engine to be run in its high efficiency region and to be downsized, so obtaining a higher average efficiency. Moreover, the engine can be turned off when the vehicle is arrested (e.g., at traffic lights) or the power request is very low (reduction of the idle losses). Compared with battery electric vehicles (BEVs), they have longer driving ranges with smaller (and lighter) batteries and the same possibility of grid recharging in the case of plug-in hybrid electric vehicles (PHEVs). BEVs, HEVs, and PHEVs have also the capability of partially recovering energy from brakes by inverting the energy flow from batteries to wheels through the electric machine.

This book on hybrid electric vehicles brings out six chapters on some of the research activities through the wide range of current issues on hybrid electric vehicles: choice of the best architecture, performance of converters and storage systems, energy management strategies, effect of specific driving conditions, power electronics, etc. Actually, these aspects can be hardly separated; in fact, they must be addressed together in order to fully exploit the potentiality of HEVs. Nevertheless, the contributions to this book have been separated in two sections. The first section deals with two interesting applications of HEVs, namely, urban buses and heavy duty working machines. The second one groups papers related to the optimization of the electricity flows in a hybrid electric vehicle, starting from the optimization of recharge in PHEVs through advance storage systems, new motor technologies, and integrated starter-alternator technologies.

A comprehensive analysis of the technologies used in HEVs is beyond the aim of the book. However, the content of this volume can be useful to scientists and students to broaden their knowledge of technologies and application of hybrid electric vehicles.

I would like to thank InTech Publisher for inviting me to be the editor of this volume, Ms. Martina Usljebrka, and the whole Publishing Process Staff for their help in coordinating the reviews, editing, and printing of the book.

Teresa Donateo
Associate Professor of Fluid Machinery, Energy Systems, and Power Generation
Università del Salento, Italy

Section 1

Applications

Chapter 1

Trends and Hybridization Factor for Heavy-Duty Working Vehicles

Aurelio Somà

Additional information is available at the end of the chapter

http://dx.doi.org/10.5772/intechopen.68296

Abstract

Reducing the environmental impact of ground vehicles is one of the most important issues in modern society. Construction and agricultural vehicles contribute to pollution due to their huge power trains, which consume a large amount of petrol and produce many exhaust emissions. In this study, several recently proposed hybrid electric architectures of heavy-duty working vehicles are presented and described. Producers have recently shown considerable attention to similar research, which, however, are still at the initial stages of development. In addition, despite having some similarities with the automotive field, the working machine sector has technical features that require specific studies and the development of specific solutions. In this work, the advantages and disadvantages of hybrid electric solutions are pointed out, focusing on the greater electromechanical complexity of the machines and their components. A specific hybridization factor for working vehicles is introduced, taking into account both the driving and the loading requirements in order to classify and compare the different hybrid solutions.

Keywords: hybrid, electric driveline, working vehicle, hybridization factor

1. Introduction

Over the past decades, the efficiency of vehicles has become a highly discussed topic due to pollution regulation requirements. Modern internal combustion engines (ICEs) have already reached remarkable performances compared with the engines of the early 1990s. However, they are still unable to consistently reverse the growth trend in pollutant emissions because the number of vehicles is also constantly increasing [1, 2]. The European Union first introduced mandatory CO_2 standards for new passenger cars in 2009 [3] and set a 2020-onward target average emission of 95 g CO_2/km for new car fleets. The automotive industry devotes

INTECH
open science | open minds

© 2017 The Author(s). Licensee InTech. This chapter is distributed under the terms of the Creative Commons Attribution License (http://creativecommons.org/licenses/by/3.0), which permits unrestricted use, distribution, and reproduction in any medium, provided the original work is properly cited.

considerable research efforts toward reducing emissions and fossil fuel dependency without sacrificing vehicle performance. Recently, manufacturers developed technologies to reduce the NOx and particulate emissions of diesel engines, such as selective catalytic reduction and diesel oxidation catalyst [4, 5]. Moreover, common rail fuel injection has led to higher-efficiency diesel engines [6, 7]. Partial substitution of fossil diesel fuel with biodiesel is an appealing option to reduce CO_2 emissions [8, 9]. In the Brazilian transportation sector, the addition of biodiesel to fossil diesel fuel has been increasing since 2012 [10].

Heavy-duty construction and agricultural vehicles also have an environmental impact. In *Agricultural Industry Advanced Vehicle Technology: Benchmark Study for Reduction in Petroleum Use* [11], the current trends in increasing diesel efficiency in the farm sector are explored. **Figure 1** shows the diesel demand in the United States, highlighting that in the agricultural and construction machinery field, the demand has remained relatively constant since 1985, representing a significant portion of the total fuel consumption. Similarly to the automotive sector, considerable efforts have been dedicated in recent years toward reducing the energy consumption of construction and agricultural machines without compromising their functionality and performance, taking into account the restrictions imposed by the recent emission regulations [12, 13]. Engine calibrations have been optimized to reduce exhaust pollutants in accordance with the U.S. Environmental Protection Agency emissions tiers. This was

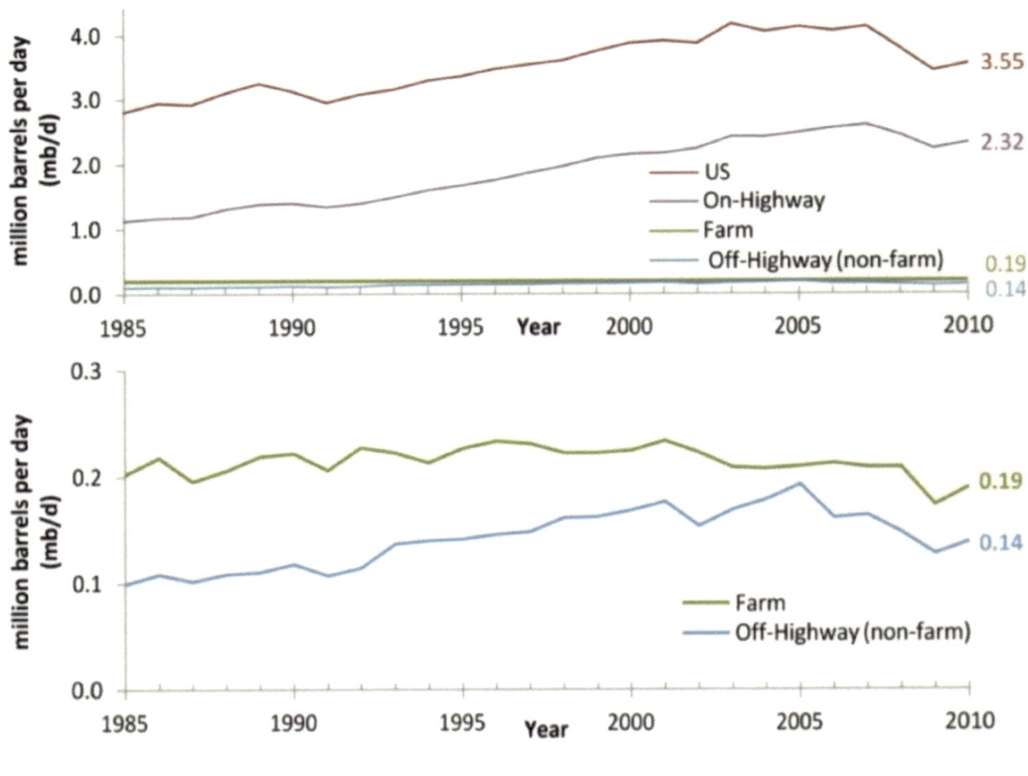

Figure 1. Historical diesel consumption in the United States. "Farm" includes agricultural diesel use; "off-highway" includes forestry, construction, and industrial use [11].

accomplished through several means, including in-cylinder combustion optimization and exhaust gas recirculation, without exhaust after-treatment systems for Tiers 1–3. With the addition of exhaust after-treatment systems for the Tier 4 interim stage, some engines require diesel exhaust fluid to catalyze pollutants in the system (e.g., urea). Some manufacturers claim as much as 5% greater fuel efficiency for their Tier 4 interim engines compared with Tier 3 models [14]; however, these entail increasing complexity, dimensions, and maintenance costs. Although most construction and agricultural vehicles include a driving mode tractor as a primary power unit, most modern models provide power for implementing a power takeoff (PTO) shaft and/or fluid power hydraulics. Moreover, working machine engines can stay idle for a notable amount of time [15]. Advanced engine controls are being introduced to reduce fuel consumption by lowering engine idle speeds and even shutting off the engine during extended idle periods. Examples of these strategies are found in existing patent applications, which indicate intentions of further development of these strategies [16]. Hybrid electric propulsion systems allow the combustion engine to operate at maximum efficiency and ensure both a considerable reduction of pollutant emissions and an appreciable decrease in energy consumption. Over the last few years, many configurations of hybrid propulsion systems have been proposed, some of which are also very complex. The fuel efficiency in this operating mode is greater than in a conventional machine for the following reasons:

- the fuel and energy consumption is limited only to the vehicle work time;
- the electronic control selects the engine speed to minimize fuel consumption depending on the state of charge of the batteries and the vehicle power demand;
- the power transmission from the electric motor to the gearbox ensures greater energy savings compared with hydraulic power transmission;
- the electric motor acts as a power unit to charge the batteries, while the vehicle is slowing down/stopping.

The automotive field has the largest number of studies, published patents, and proposals for hybrid and electric vehicles. Recently, intensive research has been carried out to find solutions that will enable the gradual replacement of the conventional engine with a highly integrated hybrid system. In the construction and agricultural working machines field, the number of concepts is limited and sporadic, and only recently has the market shown great attention to these studies. Thus, hybrid architectures allow the development of work machines characterized by high versatility and new features. Such machines can be used both indoors and outdoors because they can operate in both full electric and hybrid modes. The advantages to end users are reduction of running cost due to greater fuel efficiency and use of electric energy, and better work conditions due to low noise emissions.

From a system engineering point of view, the different solutions are described by introducing a specific hybridization factor suitable for work vehicles that include two main functionalities: driving and loading. The high-voltage electrification of work vehicles is also currently under development [17, 18]. According to Ponomarev et al. [19], in order to be competitive, manufacturers should offer energy-efficient and reliable hybrid vehicles to their customers. Compared with automobiles, the introduction of electric drives in work vehicles would allow expanded functionalities because these machines have a large variety of functional drives [20]. The first part of this report gives an overview of the components of the electrification solution and hybrid/electric architectures, discussing the advantages related to the different solutions. The machines are then schematically described and compared, showing the hybrid architectures of the proposed solutions. Finally, the introduction of a specific hybridization factor is proposed as a first classification of the main hybrid work vehicles [21, 22].

2. HEV power train configurations

The SAE defines a hybrid vehicle as a system with two or more energy storage devices, which must provide propulsion power either together or independently [23]. Moreover, an HEV is defined as a road vehicle that can draw propulsion energy from the following sources of stored energy: a conventional fuel system and a rechargeable energy storage system (RESS) that can be recharged by an electric machine (which can work as a generator), an external electric energy source, or both. The expression "conventional fuel" in the SAE definition constrains the term HEV to vehicles with a spark-ignition or a compression-ignition engine as the primary energy source. However, the United Nations definition of HEV [24] mentions consumable instead of conventional fuel. On this basis, the primary energy source in an HEV is not necessarily the engine hydrocarbon fuel, or biofuels but can also be the hydrogen fuel cell. The term electric-drive vehicle (EDV) is used in Ref. [25] to define any vehicle in which wheels are driven by an electric motor powered either by a RESS alone or by a RESS in combination with an engine or a fuel cell. Some types of EDV belong to the subset of plug-in electric vehicles (PEVs) [25, 26].

Compared with conventional internal combustion engine vehicles, HEVs include more electrical components, such as electric machines, power electronics, electronic continuously variable transmissions, and advanced energy storage devices [27]. The number of possible hybrid topologies is very large, considering the combinations of electric machines, gearboxes, and clutches, among others. The two main solutions, series and parallel hybrid, can be combined to obtain more complex and optimized architectures. There is no standard solution for the optimal size ratio of the internal combustion engine and the electric system, and the best choice includes complex trade-offs between the power as well as between cost and performance [28]. The power train configuration of an HEV can be divided into three types: series, parallel, and a combination of the two [29].

2.1. Series hybrid electric vehicles

Series hybrid electric vehicles (SHEVs) involve an internal combustion engine (ICE), generator, battery packs, capacitors and electric motors as shown in **Figure 2** [30–32]. SHEVs have no mechanical connections between the ICE and the wheels. The ICE is turned off when the battery packs feed the system in urban driving. A significant amount of energy is supplied from the regenerative braking. Therefore, the engine operates at its maximum efficiency point, leading to improved fuel efficiency and lesser carbon emission compared with other vehicle

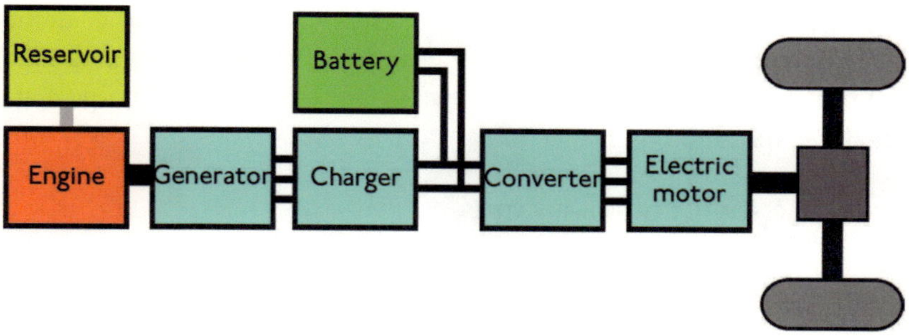

Figure 2. Schematic of series hybrid electric vehicles (SHEV).

configurations [33]. The series hybrid configuration is mostly used in heavy vehicles, military vehicles, and buses [34]. An advantage of this topology is that the ICE can be turned off when the vehicle is driving in a zero-emission zone. Moreover, the ICE and the electric machine are not mechanically coupled; thus, they can be mounted in different positions on the vehicle layout drive system [35].

2.2. Parallel hybrid electric vehicles (PHEV)

In a PHEV, mechanical and electrical powers are both connected to the driveline, as shown in **Figure 3**. In the case of parallel architectures, good performance during acceleration is possible because of the combined power from both engines [35]. Different control strategies are used

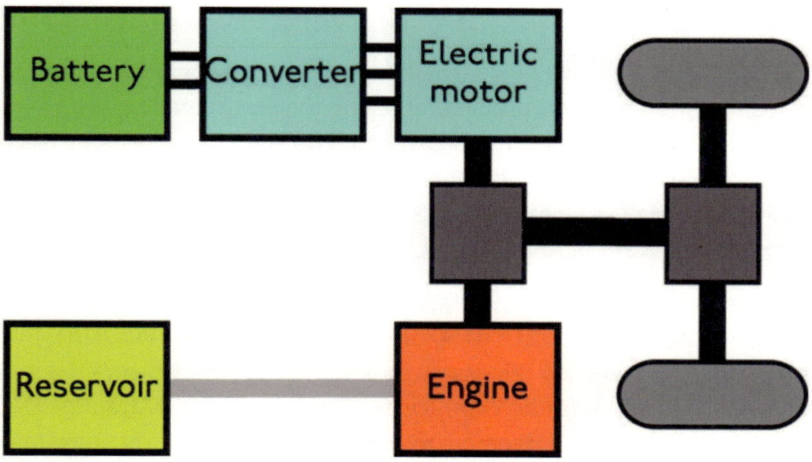

Figure 3. Schematic of parallel hybrid electric vehicles (PHEV).

in a preferred approach. If the power required by the transmission is higher than the output power of the ICE, the electric motor is turned on so that both engines can supply power to the transmission. If the power required by the transmission is less than the output power of the ICE, the remaining power is used to charge the battery packs [36]. Moreover, mechanical and electric power could be decoupled, and the system has a high operating flexibility enabling three modes of operation: purely combustion; purely electric and hybrid. Usually, a PHEVs are managed in purely electric mode at low speeds, until the battery charge state reaches a predetermined low threshold, typically 30%.

2.3. Combination of parallel and series HEVs

In the series-parallel hybrid configuration can be highlighted two main power paths. In mechanical power path, the energy generated by the combustion engine is directly transmitted to the wheels, while the electric path the energy generated by the thermal engine is converted first into electrical energy by means of the generator and then again converted to mechanical energy delivered at the wheels. It is possible therefore to have mixed architectures denominated "power splits" in which the installed power is divided by means of mechanical couplers. Combination of parallel and series hybrid configurations is further divided into subcategories based on how the power is distributed [37]. PHEVs are even more suitable topologies than HEVs for reducing fuel consumption because, unlike HEVs, they may be charged from external electric power sources [38]. In all the configurations, regenerative braking can be used to charge the battery [36]. Moreover to make recharging of batteries easier, some configurations are equipped with an on-board charger and defined Plug-in electric vehicle (PEV) [39].

3. Sub-system components of hybrid vehicles

3.1. Electric motors

The energy efficiency of a vehicle power train depends on, among other features, the size of its components. The optimization problem of sizing the electric motor, engine, and battery pack must consider both performance and cost specifications [40, 41]. Among electric motors, although the permanent magnet synchronous motor is considered as the benchmark, other types of motors are being explored for use in HEVs. Currently, there is some concern on the supply and cost of rare-earth permanent magnets.

Considerable research efforts have been made to find alternative electric motor solutions with the lowest possible use of these materials [42, 43]. For instance, some automotive applications use induction motors or switched reluctance motors [34]. **Figure 4** shows the most conceivable electric motor scenario in forthcoming years. Compared with hydraulics, electric drives provide better controllability and dynamic response and require less maintenance. Similarly to electric power, hydraulic power can be distributed quite easily on the implement; however, hydraulics suffers from poor efficiency in part-load operating conditions [44]. The specific electric drives for agricultural tractors are listed in Refs. [45, 46].

3.2. Continuous variable transmission (CVT)

Working vehicles drive at low speed, and the energy consumed in accelerating and climbing slopes should be partially recovered at decelerating and descending slopes. Compared with urban and on-road vehicles, construction and agricultural are used in a lower range of

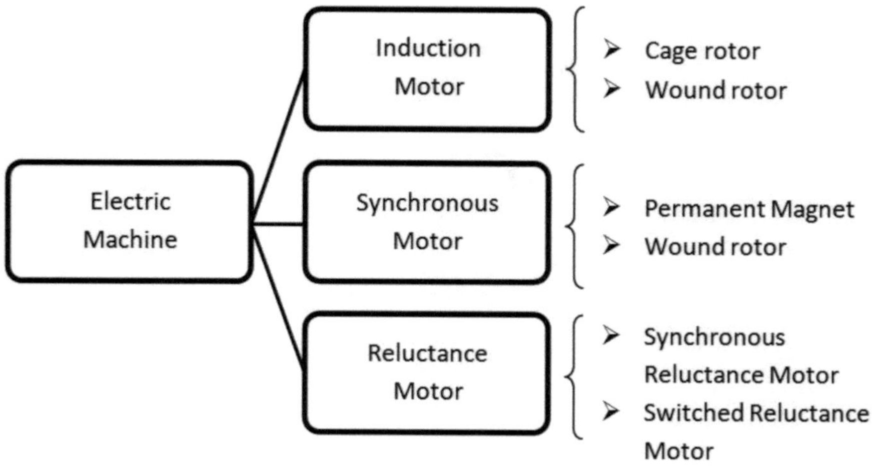

Figure 4. Types of electric motors for HEV applications.

velocity. Rolling requirements in construction and agricultural machines are related to the resistance due to tire deformation combined with resistance due to soil deformation [47, 48]. In the case of work vehicles, continuous variable transmission CVT could be used to determine the energy flow that reaches the transmission from each energy source (engine, generator, and motor battery) [49].

3.3. Energy storage devices

The energy efficiency of construction machinery is generally relatively low, and kinetic or potential energy is lost during operation [50]. Currently, batteries [51], super-capacitors, hydraulic accumulators, and flywheels are mainly used as energy storage devices in hybrid construction and agricultural machinery (HCAM), as schematically described in **Figure 5**.

3.3.1. Batteries

Batteries are the most studied energy storage and are divided into three types: Li-ion [52], nickel-metal hydride [53, 54], and lead-acid [55]. Li-ion batteries are considered as a highly prospective technology for vehicle applications [56, 57] because of their larger storage capacity, wide operating temperature range, better material availability, lesser environmental impact, safety [58–60]. However, despite having the highest energy density, Li-ion batteries a shorter lifetime, higher vulnerability to environmental temperature, and higher cost compared with other energy storage devices. A comprehensive review examined the electrochemical basis for the deterioration of batteries used in HEV applications and carried out tests on xEVs, automotive cells, and battery packs [61, 62] regarding their specific energy, efficiency, self-discharge, charge-discharge cycles, and cost. The results indicated that Li-ion is currently the best battery solution, surpassing the other technologies in all parameters except charge speed, in which Pb-acid batteries showed a better performance. Over the last years, graphene and its applications have become an important factor in improving the performance of batteries [63].

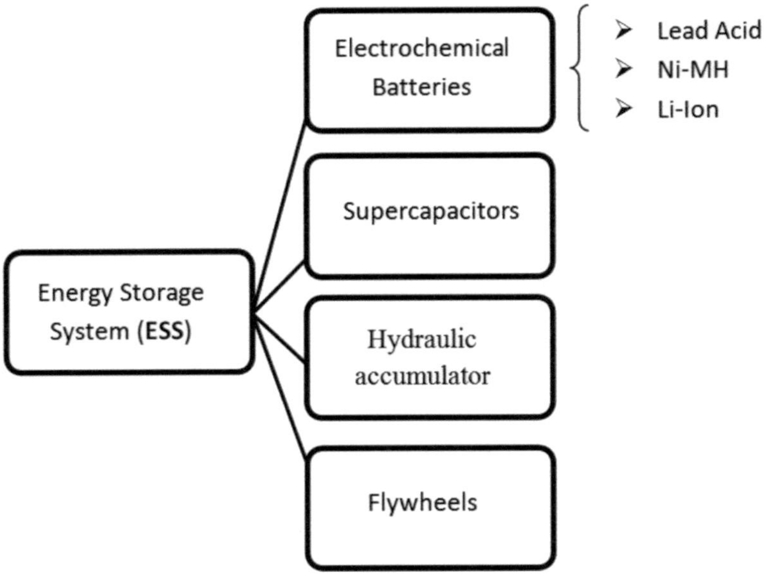

Figure 5. Energy storage for hybrid construction and agricultural machinery.

3.3.2. Supercapacitors

An alternative energy storage device for hybrid power trains could be super-capacitors, which are designed to achieve fast-charging devices of intermediate specific energy [64]. A super-capacitor [65, 66] has the advantage of a fast charge-discharge capacity, allowing a higher regenerative braking energy and supplying power for larger acceleration [67] and can be classified as a double-layer capacitor or a pseudo-capacitor according to the charge storage mode. However, the main drawback of a super-capacitor is that it has low energy density, which leads to a limited energy capacity.

3.3.3. Hydraulic accumulator

The hydraulic storage approach converts the recoverable energy into hydraulic form inside an accumulator and then releases it by using secondary components or auxiliary cylinders [68–70]. Compared with an electric hybrid system composed of a battery or super-capacitor, a hydraulic accumulator device has an advantage in power density over an electric system. Moreover, hydraulic accumulator energy recovery systems are ideal for cases of frequent and short start-stop cycles [71, 72]. However, the application of such systems in work vehicles still presents several defects: The impact of the limited energy density is a design trade-off between the energy storage capacity and volume or weight [73].

3.3.4. Flywheel energy storage system

The flywheel energy storage system (FESS) has improved considerably in recent years because of the development of lightweight carbon fiber materials. This system has become one of the most common mechanical energy storage systems for hybrid vehicles [74, 75]. When in charge mode, the electric motor drives the flywheel to rotate and store a large amount of kinetic energy (mechanical energy); when in discharge mode, the flywheel drives the generator, converting kinetic energy into electric energy [76]. The FESS has the advantages of high energy density and high power density [77] and works best at low speeds and in frequent stop-start work conditions. Producing this system could be cheaper than producing batteries; however, the system has limited storage time, and a significant percentage of the stored capacity is wasted through self-discharge [78].

4. Hybridization factor

In HEV engineering, the integration of engines, mechanical components, and electric power trains leads to increased energy efficiency, that is, a reduction in fuel consumption and a subsequent decrease in CO_2 emissions. In the automotive industry, the basic logic of a hybrid vehicle is to provide a new source of power that intervenes in place of the primary source (ICE) to improve the overall performance of the system. Moreover, there are possible modes of operation that are not provided in a conventional vehicle, such as regenerative braking and electric mode (EV). Below are some of the main advantages of a hybrid configuration over a vehicle equipped with a combustion engine alone.

- Electric motor can act both as an engine and as a generator, allowing a reversible flow of power from the battery to the wheels and vice versa.
- During braking, some of the kinetic energy is recovered (regenerative braking).
- The vehicle can be used only in the electric mode (zero emission vehicle—ZEV).
- When the vehicle has to stop temporarily, the combustion engine can be switched off, therefore ensuring considerable energy saving.

It should first be mentioned that there is actually no real classification for hybrid vehicles, although a first orientation phase can be identified by defining a significant hybridization factor (HF) as the ratio between the power of the installed electric motor and the total amount of power delivered by the combustion engine and electric motor on the vehicle:

$$HF = \frac{P_{em}}{P_{em} + P_{ICE}} \tag{1}$$

where P_{em} is the electric motor drive power, and P_{ICE} is the internal combustion engine power. In the case of conventional vehicles, the hybridization factor is clearly equal to zero, whereas in the case of electric vehicles, the hybridization factor has a unit value. Between these values, all possible solutions can be obtained. In the automotive engineering field, the definition of the hybridization factor has been extensively studied for several applications [49, 79, 80], considering its effect on performance and optimization [81–83]. Furthermore, depending on the degree of hybridization and the capacity of the hybrid propulsion system to store energy, three different levels of hybridization are defined.

- **Full hybrid** is when the electric system alone is able to make the vehicle move on a standard driving cycle ($0.5 < HF < 0.7$).

- **Mild hybrid** is when the purely electric operation mode is not able to follow a full standard driving cycle (0.25 < HF < 0.5).
- **Minimal hybrid** is equipped with a stop and start function, characterized by a decreasing distance in the purely electric mode (0 < HF < 0.1).

HF = 0 is applicable to a conventional engine vehicle, whereas HF = 1 is applicable to a "pure" electric vehicle, such as the BEV [43]. **Table 1** presents the hybridization factors calculated by using Eq. (1), taking into account the electrical driveline for automotive applications.

Compared with cars, the introduction of electric drives in tractors would allow expanded functionalities, considering that agricultural machines have a large variety of functional loading and working drives [20, 84]. The working cycle of a vehicle is strongly correlated with the application. In the case of a car, the comparison can be carried out by evaluating the extra-urban cycle and the urban cycle. For example, in the case of the urban cycle, the vehicle recovers energy due to frequent accelerations and stops. Working machines even with repetitive movements, such as excavators, are able to recover the kinetic energy of the arm. For agricultural tractors and machinery, two tasks [85] have been identified, such as working conditions with steps at which energy recovery is possible: transport and front loading. Telescopic handlers also have a similar duty cycle. Unlike in hybrid cars, the hybrid propulsion system in heavy-duty machinery can supply power to the driveline and loading hydraulic circuit [86]. The mechanical power supplied by the ICE flows to recharge the battery pack, actuate the hydraulic pump, and move the driveline (**Table 2**).

Although there is no classification for hybrid heavy-duty machines in the literature, a first orientation phase can be determined by defining a hybridization factor for a work vehicle HF_{WV} [87].

4.1. Driveline power

Hybrid architecture in series or in parallel has, in both cases, at least one electric motor (EM_1) for moving the vehicle. In order to generalize the different configurations define (EM_1), the electric motor used for the traction of the vehicle. Therefore, according to the hybridization factor described in the automotive field, the first term (μ_1) of the hybridization factor for heavy-duty vehicles (HF_{HDV}) is as follows:

$$\mu_1 = \frac{P_{em_1}}{P_{em_1} + P_{ICE}} \tag{2}$$

Vehicle	Electric motor (kW)	ICE (kW)	HF Eq. (1)
Toyota Prius	31	43	0.42
Toyota Prius 3rd gen.	50	53	0.49
Honda Insight	10	50	0.17
Honda Civic	10	63	0.07

Table 1. HF comparison among automotive vehicles [80].

	Hybrid architecture		
	Series	**Parallel**	**Series-parallel**
Driveline powered by ICE	No	Yes	Yes
Driveline powered by EM	Yes	Yes	Yes
Loading devices powered by EM	No	Yes	Yes
Loading devices powered by EG	Yes	No	Yes
Loading devices powered by ICE	No	Yes	Yes

Table 2. Architectures of hybrid construction and agricultural machinery (HCAM).

4.2. Loading power

The driveline architecture in work vehicles can be electrical, hydraulic, and/or mechanical. Moreover, the loading power can be hydraulic or electrohydraulic depending on the vehicle topology architecture. Many work machines have some hydraulic actuators to be controlled, a big difference between a passenger car and a heavy-duty vehicle. In a full hybrid vehicle, for example, the hydraulic power for loading the bucket is supplied by the hydraulic pump, which can be powered by the ICE or an electric motor (EM_2). The second ratio (μ_2) of the hybridization factor for heavy-duty vehicles can therefore be defined as follows:

$$\mu_2 = \frac{P_{em_2}}{P_{em_2} + P_{ICE}} \tag{3}$$

In the automotive industry, the power of the internal combustion engine is mainly used for the handling of the vehicle, and other functions (such as air conditioning) may be neglected in a first order assessment hybridization. In a work machine, the power of the internal combustion engine can be used for both driving operations for loading activities. In particular, it is observed that the power required to move loads or to carry out excavation work is of the same order of magnitude of power required to move the vehicle. So, the design of a hybrid working vehicle must take into account the power requirements of the working cycle with particular reference to the types of equipment that can be connected to the arm or blade of the machine. In the present work, in order to define a hybridization factor that allows comparing the many hybrid applications in the construction and agricultural machinery sector is the hypothesis that the power can be conventionally comparable between driving and loading is used.

According to the previous statement and combining the two ratios expressed in Eqs. (2) and (3), the hybridization factor for heavy-duty work vehicles can be defined as follows:

$$HF_{WV} = \frac{1}{2}(\mu_1 + \mu_2) \tag{4}$$

5. Architecture review of hybrid construction and agricultural machinery

Manufacturers, governments, and researchers have been paying increasing attention to hybrid power train technology toward decreasing the high fuel consumption rate of construction machinery [17]. Hybrid wheel loaders, excavators, and telehandlers have particularly shown significant progress in this regard [88, 89]. With hybrid work vehicles attracting more attention, power train configurations, energy management strategies, and energy storage devices have also been increasingly reported in the literature [73, 90–92]. Both researchers and manufacturers have approached studies of the hybrid power system applications, energy regeneration systems, and architectural challenges of construction machinery qualitatively but not systematically and quantitatively. A first review of an electric hybrid HCM was presented in 2010 [107]. More recently, a specific review of a wheel loader and an excavator [108] was carried out, and another work in the field of high-voltage hybrid electric tractors [109] was published. Hitachi successfully launched the first hybrid loader in 2003 [90], and Komatsu developed the first commercial hybrid excavator in 2008 [93]. Komatsu developed the HB205-1 and HB215LC-1 hybrid electric excavators, which are capable of recovering energy during the excavator slewing motion and of storing this energy in ultra-capacitors. Earth-moving machinery manufacturers have developed some diesel-electric or even hybrid-electric models . Johnson et al. [96] compared the emissions of a Caterpillar D7E diesel-electric bulldozer with its conventional counterpart [95]. Over the last years, there has been increasing interest in tractor and agricultural machinery electrification [96–99]. A number of tractor and agricultural machinery manufacturers have developed some diesel-electric or even hybrid-electric prototypes [20, 49, 100–102]. Recently, the Agricultural Industry Electronics Foundation started working on a standard for compatible electric power interfacing between agricultural tractors and implements [103], including, among others, the John Deere 7430/7530 E-Premium and 6210RE electric tractors [104] and the Belarus 3023 diesel-electric tractor [105]. Among telehandler vehicles, the TF 40.7 Hybrid telescopic handler proposed by Merlo [106]. Thus, it is necessary to study the various types of power train configurations of hybrid wheel loaders and excavators to better understand their construction features. The power requirement has different working cycles depending on the applications. Many construction machinery manufacturers and researchers have studied hybrid wheel loaders to effectively use the braking energy and operate the engine within its high-efficiency range [110–113]. According to the classification of hybrid vehicles in the automotive field, there are three main design options for hybrid wheel loader power trains: series, parallel, and series-parallel. In the literature review, the proposed architecture is mainly described, but no attempt at classification and comparison is made. It is not easy to find data sheets on the different vehicles because most of them are still at the prototype level. The comparison first outlines the architectures of the hybrid work vehicle solutions developed by the main manufacturers, as shown in **Table 3**.

Figure 6 shows the series hybrid configuration of a wheel loader. As in the configuration of a hybrid vehicle, classic engine series ICE directly drives the electric generator, the electricity so generated is used to control the electric motor connected to the driveline. The advantage of a series hybrid wheel loader is the greater simplicity. In addition the engine ICE, being decoupled from the wheels, it can be used at a fixed point in the conditions of greater efficiency.

http://dx.doi.org/10.5772/intechopen.68296

Model	Type of vehicle	Driveline	Loading and working system	Energy storage
Caterpillar—D7E	Dozer	Series	Conventional	None
Volvo—L220F Hybrid	Wheel loader	Parallel	Parallel	Battery
Mecalac—12MTX Hybrid	Articulated loader	Paralell	Paralel	Battery
John Deere—644K Hybrid	Wheel loader	Series	Parallel	None
Merlo—TF 40.7 Hybrid	Telehandler	Series	Parallel	Battery
Claas—6030 Hybrid	Telehandler	Parallel	Conventional	Battery
Kobelco SK200H Hybrid	Excavator	Conventional	Series	Battery/capacitors
Komatsu—HB215 LC-1	Excavator	Paralel	Parallel	Supercapacitors

Table 3. Hybrid working vehicles and their architectures.

In the case of hybrid wheel loaders in series from the transformation of mechanical power into electrical and drive of the electric motor can also be done with a battery pack reduced but the generator and the electric motor need to be manufactured in terms of maximum power demand. The presence of the battery pack can allow to better manage the power demand peaks without the need to over-dimension the motor ICE [114, 115]. In literature, the hybrid drive train in the series has been applied mainly in large tonnage hybrid wheel loader.

In 2009, Caterpillar came out with the first electric hybrid bulldozers. The Caterpillar D7E model is within the range of medium dozers and replaced the traditional model D7R [94].

The company claimed an increase of productivity and a reduction in fuel consumption up to 24% over the conventional model [94]. The driveline architecture is of the series electric hybrid type, as described in **Figure 6**, with the electric motors powered directly from the inverter but having the peculiarity to be directly charged from the ICE without any accumulation system.

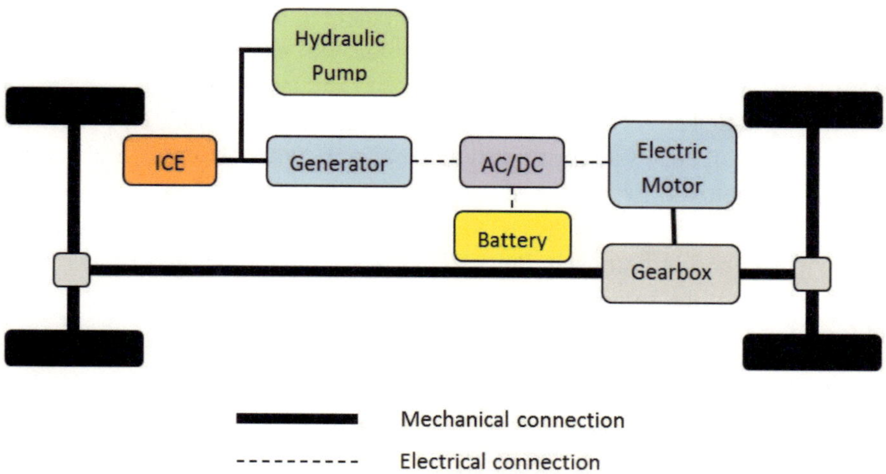

Figure 6. Working vehicle with series hybrid configuration.

The hydraulic system has a conventional architecture. **Table 4** shows the main parameters of this work vehicle. A parallel hybrid power train configuration has two separate power sources that can directly power the loader. The disadvantage of a parallel configuration is that the engine cannot always be controlled in its high-efficiency operating region because it is still mechanically coupled to the wheels with an increased efficiency compared with the conventional model and a fuel consumption reduction of 10%[116]. **Figure 7** shows a schematic of the Volvo L220F parallel hybrid electric wheel loader (HEWL). The vehicle has a parallel hybrid electric architecture for both the driving and the loading system. The basic idea of this parallel hybrid layout is to supply additional electric power when necessary, regenerating the machine during normal operations and minimizing the consumption in idle conditions. The power required by the device can be flexibly provided by using a work pump, which is driven by the pump motor shows the main parameters of the Volvo L220F. Mecalac proposed a similar architecture for the 12 MTX hybrid model and claimed to save up to 20% in fuel consumption [117].

However, the parallel configuration is still on the researching stage, and Liugong has applied a solution with super-capacitors instead of batteries [118] as schematically shown in **Figure 8**.

At the CONEXPO International Trade Fair for Construction Machinery (2011), John Deere presented the first prototype of its hybrid wheel loader, the 944K hybrid. In February 2013, the entry of the first hybrid wheel loader, the 644K hybrid, in the market was announced with a reduction in fuel consumption up to 25% [119]. In this smaller model, a single electric machine provides all the power needed to drive the vehicle. The vehicle driveline has a series electric hybrid architecture, with the electric motor directly powered by the inverter without an energy storage system. **Figure 9** shows a schematic view of John Deere 644K hybrid wheel loader [120]. The installed electrical machines are liquid-cooled brushless permanent magnet motors.

The innovative architecture proposed by Merlo, as shown in **Figures 10** and **11**, is considered as a fully series architecture for vehicle traction and as a parallel architecture for the operation of hydraulic systems. This kind of innovative, patented series-parallel architecture, with a split input for hydraulic lifting, allows both the electrical and the mechanical components to be arranged in a way that is compatible with the current layout and performance of Merlo machines. The main objectives of this hybrid telehandler are an overall improvement in performance, a decrease in daily fuel consumption in ordinary work activities, and a reduction in noise emissions. Moreover, the proposed configuration is capable of working in full electric, zero-emission mode for indoor use, such as in cattle sheds, stables, industrial and food processing warehouses, and buildings. In Ref. [87], it has been demonstrated a fuel consumption reduction of 30% with the same level of dynamic performance compared with the conventional telehandler.

Claas proposed a parallel mild hybrid solution for the Scorpion telehandler. The simulation results reported in Refs. [121, 122] show a reduction in fuel consumption of about 20% and emissions for this parallel hybrid solution compared with the traditional model. The solution proposes the use of the electric motor as a power boost to maintain the performance while using a smaller diesel motor.

Vehicle	Vehicle operative weight (t)	Electric driving motor (P_{EM1}) (kW)	Electric loading motor (P_{EM2}) (kW)	ICE (kW)	HF (Eq. 2)	HF (Eq. 3)	HF_{WV} (Eq. 4)	% of fuel reduc
Caterpillar D7E	26	2*60	0	176	0.40	0	0.20	24
Volvo—L220F Hybrid	32	50	50	259	0.16	0.16	0.16	10
Mecalac—12MTX Hybrid	8.3	20	20	51	0.28	0.28	0.28	20
John Deere—644K Hybrid	19	80	80	171	0.32	0.32	0.32	25
Merlo—TF 40.7 Hybrid	7.5	60	40	56	0.52	0.42	0.47	30
Claas—6030 Hybrid	5.6	15	0	55	0.21	0	0.11	20
Kobelco SK200H Hybrid	20	0	37	114	0	0.24	0.12	40
Komatsu—HB215 LC-1	21	20	90	104	0.16	0.46	0.31	25

Table 4. List of hybrid working vehicles HF and claimed fuel reduction.

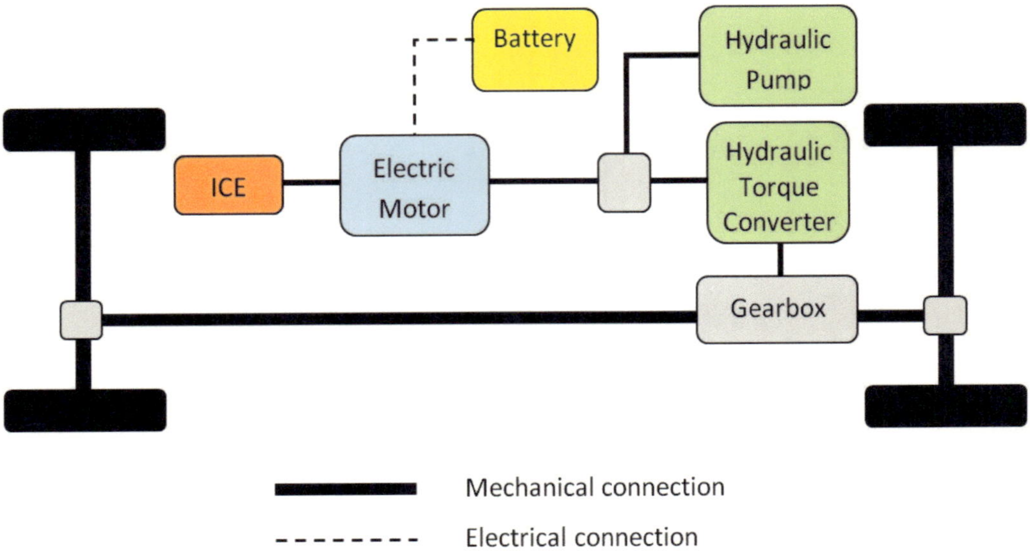

Figure 7. Parallel hybrid configuration of the Volvo L220F hybrid [116].

The excavator is a type of construction machinery with a larger weight and higher energy consumption [107]. A hybrid excavator can typically recycle two energy types, including the braking kinetic energy of the swing and the gravitational potential energy of the booms. In the recent literature, excavators present a wide combination of series, parallel, or series-parallel hybrid architectures. The change in configuration and the additional costs of electrical components make the commercialization of hybrid configurations difficult. **Figure 12** shows the

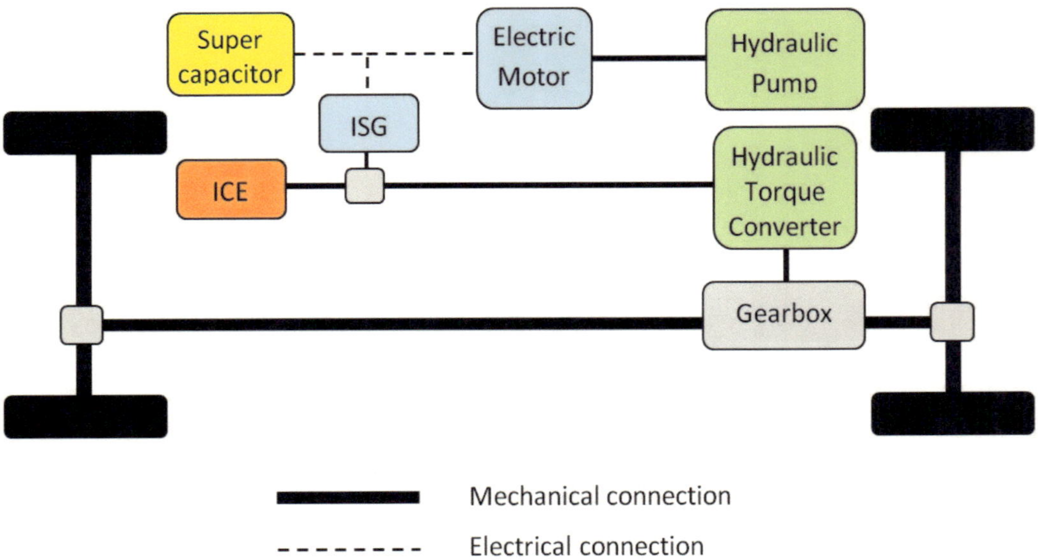

Figure 8. Parallel hybrid configuration with super-capacitors, as applied by Liugong [118].

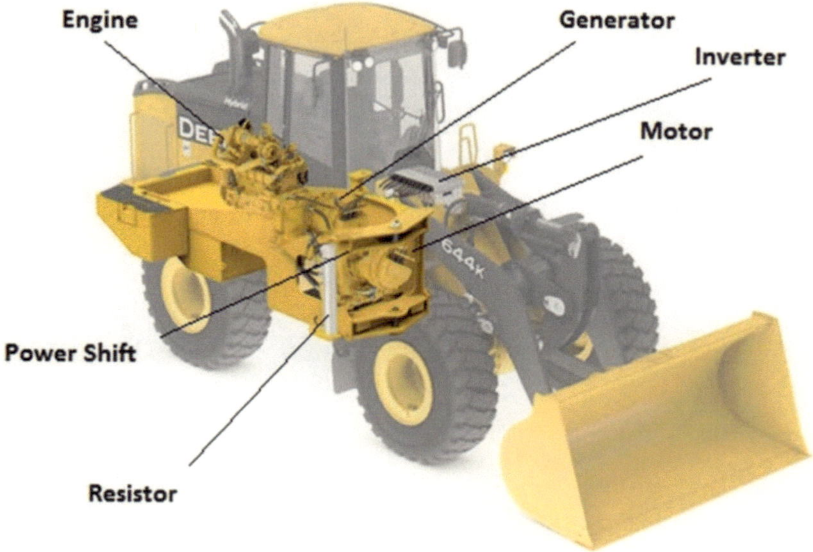

Figure 9. Schematic of the John Deere 644K hybrid wheel loader [120].

schematic of the Kobelco series hybrid excavator; the first prototype of this 6-t configuration was developed in 2007 with a claimed in [123] to cut fuel consumption by 40% or more and reporting results of the verification test on the efficiency of the hybrid excavator in different working cycle operations [124, 125].

As showed in **Figure 12** in the hybrid solution proposed by Kobelco, each hydraulic is driven by an electric motor. This solution increases efficiency but the production cost is higher.

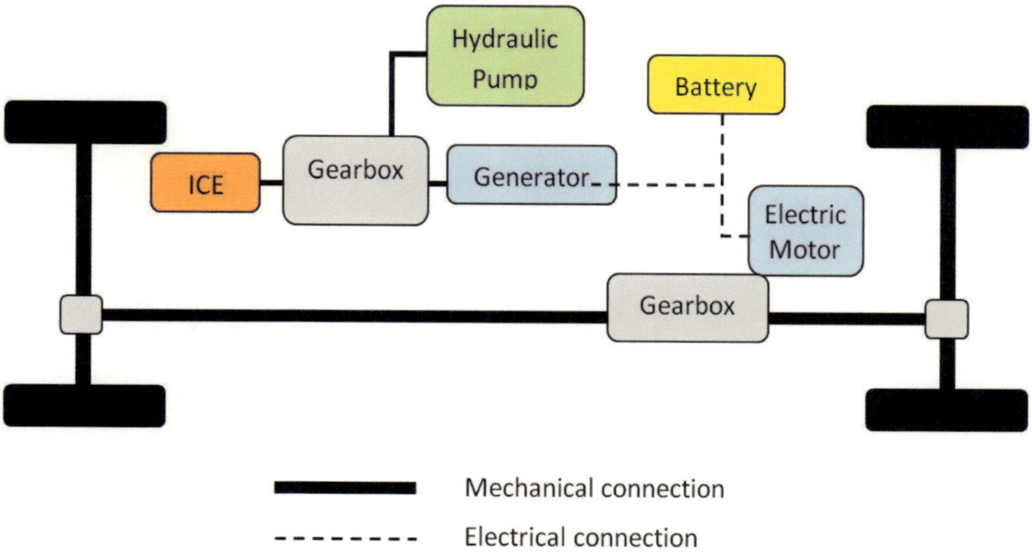

Figure 10. Series-parallel hybrid configuration of a Merlo working vehicle [106].

Figure 11. View of Merlo – TF 40.7 hybrid the hybrid telehandler [106].

In the case of parallel hybrid excavator, the internal combustion engine operates the hydraulic pump and generator. The hydraulic pump drives the hydraulic circuit of the device, in a manner similar to conventional excavators, while the generator transforms the mechanical energy into electrical power and can operate the electric motor of swing rotation. The hybrid solution in parallel is simpler; however, the fuel consumption is higher, and the return time for these working machines is longer [126]. Hitachi, as shown in **Figure 13**, proposed a parallel hybrid excavator with the gravitational potential recovery of the boom [113].

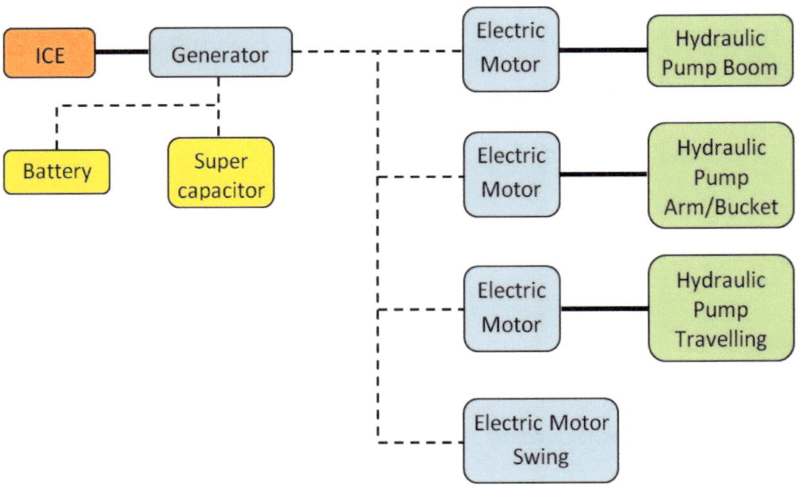

Figure 12. Series hybrid configuration of the Kobelco excavator.

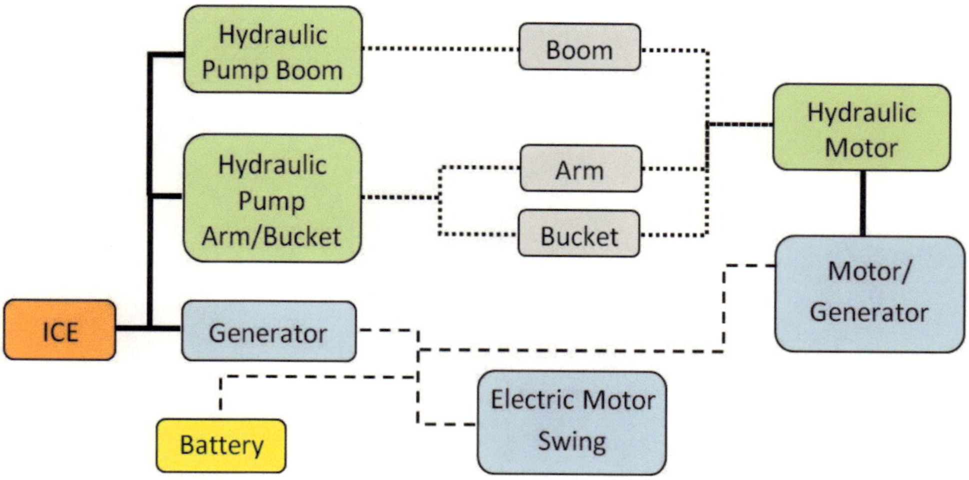

Figure 13. Parallel hybrid configuration for working excavator Hitachi [127].

In the series-parallel hybrid power train configuration of an excavator, the engine drives the generator directly. The hydraulic pumps are driven by the generator in series, and the swing electric motor is powered by the generator and the battery or super-capacitor in parallel. Although series-parallel hybrid excavators have higher production costs compared with parallel and series structures, they offer the shortest cost recovery time and efficiency with a fuel consumption up to 25% [126]. Series-parallel hybrid excavators are regarded as the most promising solutions, and both Komatsu (**Figures 14** and **15**) and Doosan use similar configurations [128, 129].

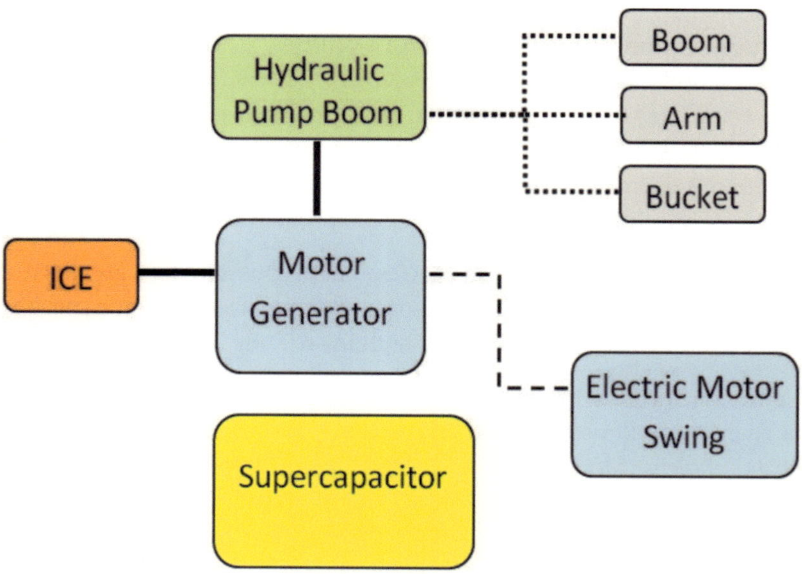

Figure 14. Series-parallel hybrid configuration for working excavator Komatsu [126].

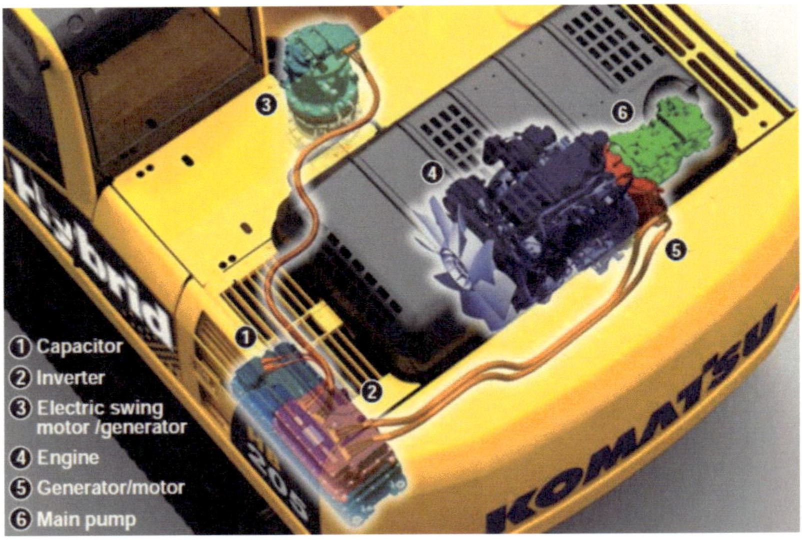

Figure 15. Schematic view of Komatsu—HB215 LC-1 hybrid excavator [126].

The attempt at classification in the present work is based on the specific HF defined in Section 4, taking into account the data sheets of the vehicles. **Table 4** shows the hybridization factors for work machines, calculated by using Eq. (4) [22] and considering the effect of a hybrid electric driveline and hybrid electric loading/working functions.

6. Trends and conclusions

This study focused on the electrification of work vehicles, such as agricultural machineries, which is still in the research and development stage. Similarly to HEVs, the main design issue in HACMs is controlling the energy transfer from the sources to the loads with minimum loss of energy, which is dependent on the driving and working cycles. Compared with automobiles, the introduction of electric drives in tractors would allow expanded functionalities because agricultural machines have a large variety of functional drives.

Main differences in requirements, working cycles, and proposed hybrid architectures between HEV and HACM were determined along the present study, focusing on a specific hybridization factor for working vehicles that consider both the driving and the loading electrification.

The hybridization factor for working vehicles is introduced in order to classify and compare the different hybrid solutions proposed by main manufactures taking into account different architectural choices. Moreover, the claimed increasing of efficiency due to the power train electrification is reported and listed in terms of fuel consumption reduction. Taking into account a large variety of architectural hybrid solution, it has been proven a good correlation between the hybridization factor and the fuel efficiency as a general trend in benefit of hybrid electrification of working machine.

Because charging a battery pack from the grid is more efficient than charging it from a tractor engine, it seems logical to hybridize the tractor with high-voltage batteries and propulsion motors. In this manner, the internal combustion engine could be downsized, and the traction battery pack could be charged from the grid. Fuel consumption costs would thus decrease. However, compared with traditional construction machinery, an additional energy storage device is needed, which increases the initial costs. Moreover, the cost added by high-voltage equipment needs to be considered in the whole turnover of the hybrid vehicle conversion. As indicated by several reports and prototypes, hybrid systems have promising applications in both agricultural and construction machinery, but major drawbacks are related to the increased cost due to electrification. Hybrid technologies, particularly energy storage devices, are still in the early stages of development, and the trends in cost reduction could push researchers and manufacturers toward the optimization of hybrid solutions for HCAM.

Abbreviation

BEV	Battery electric vehicle	CVT	Continuously variable transmission
EDV	Electric-drive vehicle	FCV	Fuel cell vehicle
HEV	Hybrid electric vehicle	SHEV	Series hybrid electric vehicle
PHEV	Parallel hybrid electric vehicle	HF	Hybridization factor
ICE	Internal combustion engine	PEV	Plug-in electric vehicle
ZEV	Zero emission vehicle	ESS	Energy storage system
RESS	Rechargeable energy storage system	BMS	Battery management system
HCAM	Hybrid construction and agricultural machineries		

Author details

Aurelio Somà

Address all correspondence to: aurelio.soma@polito.it

Department of Mechanical and Aerospace Engineering, Politecnico di Torino, Italy

References

[1] Azhar KM, Zahir KM, Zaman K, et al. Global estimates of energy consumption and greenhouse gas emissions. Renewable & Sustainable Energy Reviews. 2014;**29**:336-344

[2] Reşitoğlu A, Altinis K,Keskin A. The pollutant emissions from diesel-engine vehicles and exhaust aftertreatment systems. Clean Technologies and Environmental Policy. 2015;**17**:15-27

[3] Regulation (EC) No. 443/2009 of the European Parliament and of the Council of 23 April 2009 setting emission performance standards for new passenger cars as part of the Community's integrated approach to reduce CO_2 emissions from light-duty vehicles. Official Journal of the European Union L140/1.5/6/2009

[4] Clark NN. NOx/fuel tradeoff for powertrain technologies. In: Heavy-duty Vehicle Efficiency Technical Workshop. San Francisco, CA, USA; October 22, 2013

[5] Du J, Sun W, Guo L, Xiao S, Tan M, Li G, et al. Experimental study on fuel economies and emissions of direct-injection premixed combustion engine fueled with gasoline/diesel blends. Energy Conversion Management. 2015;**100**:300-309

[6] Xu-Guang T, Hai-Lang S, Tao Q, Zhi-Qiang F, Wen-Hui Y. The impact of common rail system's control parameters on the performance of high power diesel. Energy Procedia. 2012;**16**:2067-2072

[7] Cwikowski P, Teodrorcyck A. The latest achievements in gasoline and diesel injection technology for the internal combustion engines. Journal of KONES. Powertrain and Transport. 2009;**16**(2):79-90

[8] Pali HS, Kumar N, Alhasan Y. Performance and emission characteristics of an agricultural diesel engine fueled with blends of Salmethylesters and diesel. Energy Conversion and Management. 2015;**90**(146):153

[9] Ettl J, Thuneke K, Remmele E, Emberger P, Widmann B. Future biofuels and driving concepts for agricultural tractors. In: 22nd European Biomass Conference & Exhibition. Hamburg, Germany; 2014

[10] Flórez-Orrego D, Silva JAM, de Oliveira Jr S. Exergy and environmental comparison of the end use of vehicle fuels: The Brazilian case. Energy Conversion and Management. 2015;**100**:220-231

[11] Hoy R, Rohrer R, Liska A, Luck J, Isom L, Keshwani D. Agricultural industry advanced vehicle technology: Benchmark study for reduction in petroleum use. Idaho National Laboratory, University of Nebraska - Lincoln - USA; 2014

[12] Moya A, Barreiro P. Moya A, Barreiro P. Recortar emisiones en vehículos agrícolas-Introducción del Tier 4: camino hacia las cero emisiones en vehículos todoterreno. Tierras. 2011;**176**:88-94

[13] Fiebig M, Wiartalla A, Holderbaum B, Kiesow S. Particulate emissions from diesel engines: Correlation between engine technology and emissions. Journal of Occupational Medicine and Toxicology. 2014;**9**:6

[14] "Meeting EPA 2012 Tier 4 Interim and EU Stage IIIB Emissions Customer FAQ (75-173 hp),"2013, Bulletin 4087191, Cummins Inc., http://cumminsengines.com/uploads/docs/4087191.pdf, last accessed January 22, 2014

[15] Rahman SM, Masjuki HH, Kalam MA, Adebin MJ, Sanjid A, Sajjad H. Impact of idling on fuel consumption and exhaust emissions and available idle-reduction technologies for diesel vehicles—A review. Energy Conversion and Management. 2013;**74**:171-182

[16] Baroni M, Sereni E, Mancarella F. Engine control device for a work vehicle. 2013. Patent Application WO2013079324 A1

[17] Holt GD, Edwards DJ. Analysis of United Kingdom off-highway construction machinery market and its consumers using new-sales data. Journal of Construction Engineering and Management I ASCE. 2012;**139**:529-537

[18] Heckmann M, Gobor Z, Huber S, Kammerloher T, Bernhardt H. Design of a test bench for traction drive systems in mobile machines. Landtechnik. 2013;**68**(6):415-419

[19] Ponomarev P, Minav T, Aman R, Luostarinen L. Integrated electro-hydraulic machine with self-cooling possibilities for non-road mobile machinery. Journal of Mechanical Engineering. 2015;**61**(3):207-213

[20] Karner J, Baldinger M, Schober P, Reichl B, Prankl H. Hybrid systems for agricultural engineering. LandTechnik. 2013;**68**(1):22-25

[21] Somà A. Effects of driveline hybridization on fuel economy and dynamic performance of hybrid telescopic heavy vehicles. In: Proc Technologies for High Efficiency & Fuel Economy; 29-30 September; Rosemont (Ill USA): SAE; 2013

[22] Somà A, Bruzzese F, Mocera F, Viglietti E. Hybridization factor and performance of hybrid electric telehandler vehicle. IEEE Transactions on Industry Applications. 2016;**52**:5130-5138

[23] SAE Intl. SAE J1715. Information report, hybrid vehicle (HEV) and electric vehicle (EV) terminology; October 2014

[24] UNECE. Working party on transport statistics. Definitions of vehicle energy types. ECE/Trans/WP.6/2011/5; 2011

[25] Nemry F, Leduc G, Muñoz A. Plug-in hybrid and battery electric vehicles: State of the research and development and comparative analysis of energy and cost efficiency. JRC Tech Notes. European Commission JRC-IPTS; 2009

[26] Katrašnik Tomazˇ. Analytical framework for analyzing the energy conversion efficiency of different hybrid electric vehicle topologies. Energy Conversion and Management. 2009;**50**(8):1924-1938

[27] Emadi A, Rajashekara K, Williamson SS, Lukic SM. Topological overview of hybrid electric and fuel cell vehicular powerX system architectures and configurations. IEEE Transactions on Vehicular Technology.2005;54(3):763-770

[28] Gao L, Dougal RA, Liu S. Power enhancement of an actively controlled battery/ultracapacitor hybrid. IEEE Transactions on Power Electronics. 2003;20(1):366-373

[29] Lo EWC. Review on the configurations of hybrid electric vehicles. In: 3rd International Conference Power Electronics Systems and Applications; 2009. pp. 1-4

[30] Bailo Camara M, Gualous H, Gustin F, Berthon A. Design and new control of DC/DC converters to share energy between supercapacitors and batteries in hybrid vehicles. IEEE Transactions on Vehicular Technology. 2008;**57**(5):2721-2735

[31] Yoo H, Sul S-K, Park Y, Jeong J. System integration and power-flow management for a series hybrid electric vehicle using supercapacitors and batteries. IEEE Transactions on Industry Applications. 2008;**44**(1):108-114

[32] Gao J, Sun F, He H, Zhu GG, Strangas EG. A comparative study of supervisory control strategies for a series hybrid electric vehicle. In: Power and Energy Engineering Conference; Asia-Pacific; 2009. pp. 1-7

[33] Northcott DR, Filizadeh S, Chevrefils AR. Design of a bidirectional buck-boost dc/dc converter for a series hybrid electric vehicle using PSCAD/EMTDC. In: IEEE Vehicle Power and Propulsion Conference; 2009. pp. 1561-1566

[34] Ehsani M, Gao Yimin, Miller JM. In: Hybrid electric vehicles: Architecture and motor drives. Proceedings of the IEEE. 2007;**95**(4):719-728

[35] Kamil Çağatay Bayindir,Mehmet Ali Gözüküçük,Ahmet Teke.A comprehensive overview of hybrid electric vehicle: Powertrain configurations, powertrain control techniques and electronic control units. Energy Conversion and Management. 2011;**52**(2):1305-1313

[36] Katrašnik T, Tranc F, Rodman Oprešnik S. Analysis of the energy conversion efficiency in parallel and series hybrid powertrains. IEEE Transactions on Vehicular Technology. 2007;**56**(6/2):3649-3659

[37] Cheong K, Li P, Sedler S, Chase T. Comparison between input coupled and output coupled power-split configurations in hybrid vehicles. In: Proceedings of the 52nd National Conference on Fluid Power; Milwaukee;National Fluid Power Association; 2011

[38] He Y, Chowdhury M, Pisu P,Ma Y. An energy optimization strategy for power-split drivetrain plug-in hybrid electric vehicles. Transportation Research Part C. 2012;**22**:29-41

[39] Murgovski N, Johaneson L, Sjöberg J, Egardt B. Component sizing of a plug-in hybrid electric powertrain via convex optimization. Mechatronics. 2012;**22**:106-120

[40] Torres O, Bader B, Romeral JL, Lux G, Ortega JA. Influence of the final drive ratio, electric motor size and battery capacity on fuel consumption of a parallel plug-in hybrid electric vehicle. 19th International Conference on Urban Transport and the Environment; WIT Press; 2013

[41] Ebbesen S, Elbert P, Guzzella L. Engine downsizing and electric hybridization under consideration of cost and drivability. Oil and Gas Science Technology – Revued'IFP Energies Nouvelles. 2013;**68**(1):109-116

[42] Kiyota K, Kakishima T, Chiba A. Comparison of test result and design stage prediction of switched reluctance motor competitive with 60-kW rare-earthv permanent magnet motor. IEEE Transactions on Industrial Electronics.2014;**61**(10):5712-5721

[43] Boldea I, Tutelea LN, Parsa L, Dorrell D. Automotive electric propulsion systems with reduced or no permanent magnets: An overview. IEEE Transactions on Industrial Electronics.2014;**61**(10):5696-5711

[44] Karner J, Baldinger M, Reichl B. Prospects of hybrid systems on agricultural machinery. GSTF Journal on Agricultural Engineering.2014;**1**(1):33-37

[45] Bernhard B. Hybrid drives for off-road vehicles. In: FISITA World Automotive Congress; Barcelona, Spain; 2004

[46] Karner J, Prankl H, Kogler F. Electric drives in agricultural machinery. In: CIGR Ag Eng 2012; Valencia, Spain; 2012

[47] Osinenko PV, Geisler M, Herlitzius T. A method of optimal traction control for farm tractors with feedback of drive torque. Biosystems Engineering.2015;**129**:20-33

[48] Rossi C, Pontara D, Casadei D. E-CVT power split transmission for off-road hybrid electric vehicles. Vehicle Power & Propulsion Conference (VPPC) Coimbra; Portugal. IEEE; 2014

[49] Katrasnik T. Hybridization of powertrain and downsizing of IC engine—a way to reduce fuel consumption and pollutant emissions—Part I. Energy Conversion and Management. 2007;**48**:1411-1423

[50] Schoenung S. Energy storage systems cost update. Report, Sandia National Laboratories, USA; April 2011

[51] Pollet BG, Staffell I, Shang JL. Current status of hybrid, battery and fuel cell electric vehicles: From electrochemistry to market prospects. Electrochimica Acta. 2012;**84**:235-249

[52] Prada E, Domenico DD, Creff Y, et al. A simplified electrochemical and thermal aging model of LiFePO4-graphite Li-ion batteries: Power and capacity fade simulations. Journal of the Electrochemical Society. 2013;**160**:A616–A628

[53] Verbrugge MW, Conell RS. Electrochemical and thermal characterization of battery modules commensurate with electric vehicle integration. Journal of the Electrochemical Society.2002;**149**:A45–A53

[54] Kroeze RC, Krein PT. Electrical battery model for use in dynamic electric vehicle simulations. In: Power Electronics Specialists Conference; 15-19 June 2008; Rhodes, Greece. New York: IEEE; pp. 1336-1342

[55] Ceraolo M. New dynamical models of lead-acid batteries. IEEE Transactions on Power Systems. 2000;**15**:1184-1190

[56] Hu Y, Yurkovich S, Guezennec Y, et al. Electro-thermal battery model identification for automotive applications. Journal of Power Sources. 2011;**196**:449-457

[57] Chan HL. A new battery model for use with battery energy storage systems and electric vehicles power systems. Paper No. 1. In: Power Engineering Society Winter Meeting; 23-27 January 2000; Piscataway, NJ. New York: IEEE; pp.470-475

[58] Thielmann A, Sauer A, Isenmann R, et al. Produkt-roadmap lithium-ionen-batterien 2030. Report. Fraunhofer ISI, Germany; February 2012

[59] Waag W, Fleischer C, Sauer DU. Critical review of the methods for monitoring of lithium-ion batteries in electric and hybrid vehicles. Journal of Power Sources.2014;**258**:321-339

[60] Su X, Wu Q, Li J, et al. Silicon-based nanomaterials for lithium-ion batteries: A review. Advanced Energy Materials. 2014;**4**:1-23

[61] Cherry J. Battery durability in electrified vehicle applications: A review of degradation mechanisms and durability testing. FEV North America Report; 2016

[62] Mousazadeh H, Keyhani A, Javadi A, Mobli H, Abrinia K, Sharifi A. Evaluation of alternative battery technologies for a solar assist plug-in hybrid electric tractor. Transportation Research Part D.2010;**15**:507-512

[63] Kucinskis G, Bajars G, Kleperis J. Graphene in lithium ion battery cathode materials: A review. Journal of Power Sources. 2013;**240**:66-79

[64] Snook GA, Kao P, Best AS. Conducting-polymerbased supercapacitor devices and electrodes. Journal of Power Sources. 2011;**196**:1-12

[65] Zhang K, Zhang LL, Zhao XS, et al. Graphene/polyaniline nanofiber composites as supercapacitor electrodes. Chemistry of Materials. 2010;**22**:1392-1401

[66] De Souza VHR, Oliveira MM, Zarbin AJG. Thin and flexible all-solid supercapacitor prepared from novel single wall carbon nanotubes/polyaniline thin films obtained in liquid–liquid interfaces. Journal of Power Sources. 2014;**260**:34-42

[67] Peng C, Zhang S, Jewell D, et al. Carbon nanotube and conducting polymer composites for supercapacitors. Progress in Natural Science.2008;**18**:777-788

[68] Van de Ven JD. Constant pressure hydraulic energy storage through a variable area piston hydraulic accumulator. Applied Energy. 2013;**105**:262-270

[69] Liang X, Virvalo T. An energy recovery system for a hydraulic crane. Proceedings of the Institution of Mechanical Engineers, Part C: Journal of Mechanical Engineering Science.2001;**215**:737-744

[70] Ho TH, Ahn K. Design and control of a closed-loop hydraulic energy-regenerative system. Automation in Construction. 2012;**22**:444-458

[71] Wang T, Wang Q. An energy-saving pressure-compensated hydraulic system with electrical approach. IEEE/ASME Transactions on Mechatronics. 2014;**19**:570-578

[72] Wang T, Wang Q, Lin T. Improvement of boom control performance for hybrid hydraulic excavator with potential energy recovery. Automation in Construction. 2013;**30**:161-169

[73] Ho TH, Ahn KK. Modeling and simulation of hydrostatic transmission system with energy regeneration using hydraulic accumulator. Journal of Mechanical Science and Technology. 2010;**24**:1163-1175

[74] Jaafar A, Akli CR, Sareni B, et al. Sizing and energy management of a hybrid locomotive based on flywheel and accumulators. IEEE Transactions on Vehicular Technology. 2009;**58**:3947-3958

[75] Lu X, Iyer K, Mukherjee K, et al. Study of permanent magnet machine based flywheel energy storage system for peaking power series hybrid vehicle control strategy. In: Transportation Electrification Conference and Expo(ITEC); 16-19 June 2013; Dearborn, US. New York: IEEE; 2013

[76] Dhand A, Pullen K. Review of flywheel based internal combustion engine hybrid vehicles. International Journal of Automotive Technology. 2013;**14**:797-804

[77] Sebastian R, Pena Alzola R. Flywheel energy storage systems: Review and simulation for an isolated wind power system. Renewable & Sustainable Energy Reviews. 2012;**16**:6803-6813

[78] Prodromidis GN, Coutelieris FA. Simulations of economic and technical feasibility of battery and flywheel hybrid energy storage systems in autonomous projects. Renewable Energy. 2012;**39**:149-153

[79] Katrašnik T. Hybridization of powertrain and downsizing of the IC engine – Analysis and parametric study – Part 2. Energy Conversion Management. 2007;**48**(5):1424-1434

[80] Lukic SM, Emadi A. Effects of drivetrain hybridization on fuel economy and dynamic performance of parallel hybrid electric vehicles. IEEE Transactions on Vehicular Technology. 2004;**53**(2):385-389

[81] Holder C, Gover J. Optimizing the Hybridization Factor for a Parallel Hybrid Electric Small Car Vehicle Power and Propulsion Conference; VPPC '06. IEEE; 2006

[82] Bolvashenkov I, Herzog H-G, Engstle A. Factor of hybridization as a design parameter for hybrid vehicles. In: Proceedings of the IEEE International Symposium on Power Electronics, Electrical Drives, Automation and Motion (SPEEDAM); 2006. pp. 38-41

[83] Buecherl D, Bolvashenkov I, Herzgov H-G. Verification of the optimum hybridization factor as design parameter of hybrid electricvehicles. Vehicle Power and Propulsion Conference; VPPC '09. IEEE; 2009. pp. 847-851

[84] Barthel J, Gorges D, Bell M, Munch P. Energy management for hybrid electric tractors combining load point shifting, regeneration and boost. In: Vehicle Power &Propulsion Conference (VPCC); 27-30 October; Coimbra, Portugal: IEEE; 2014

[85] O'Keefe M, Simpson A, Kelly K, Pedersen D. Duty cycle characterization and evaluation towards heavy hybrid vehicle applications. SAE Tech; 2007. Paper 2007-01-0302

[86] Banjac T, Trenc F, Katrasnik T. Energy conversion efficenty of hybrid electric heavy-duty vehicles operating according to diverse drive cycles. Energy Conversion and Management. 2009;**50**:2865-2878

[87] Somà A, Bruzzese F, Viglietti E. Hybridization factor and performances of hybrid electric telescopic heavy vehicles. In: 2015 Tenth International Conference on Ecological Vehicles and Renewable Energies (EVER); 2015

[88] Ochiai M. Development for environment friendly construction machinery. Construction. 2003;**9**:24-28

[89] Inoue H. Introduction of PC200-8 hybrid hydraulic excavators. Technical Report. 2009;**54**(161):1-6. Komatsu, Japan, 27 March 2009

[90] Kanezawa Y, Daisho Y, Kawaguchi T. Increasing efficiency of construction machine by hybrid system. In: JSAE (Society of Automotive Engineers of Japan) Annual Congress; 23-25 May 2001; Yokohama, Japan. Tokyo: JSAE; Paper No. 100, 2001. pp. 17-20

[91] Lin T, Wang Q, Hu B, et al. Research on the energy regeneration systems for hybrid hydraulic excavators. Automation in Construction. 2010;**19**:1016-1026

[92] Wang D, Guan C, Pan S, et al. Performance analysis of hydraulic excavator powertrain hybridization. Automation in Construction. 2009;**18**:249-257

[93] Nishida Y. Introducing the HB335/HB365-1 hybrid hydraulic excavators. Technical Report.2014;**59**(166):1-8. Komatsu, Japan, 28 March 2014

[94] Wang H, Liu L, Zheng G, Liu X, et al. Study of two-motor hybrid bulldozer. SAE Technical Paper 2014-01-2376; 2014. DOI:10.4271/2014-01-2376

[95] Filla R. Alternative system solutions for wheel loaders and other construction equipment. In: First International Forum Alternative &Hybrid Drive Trains. Berlin, Germany; 2008

[96] Johnson KC, Burnette A, Cao T, Russell RL, Scora G. Hybrid off-road equipment in-use emissions evaluation. FY 2010-11 air quality improvement project. Hybrid off-road equipment pilot project. California Air Resources Board; 2013

[97] Tritschler PJ, Bacha S, Rullière E, Husson G. Energy management strategies for an embedded fuel cell system on agricultural vehicles. In: XIX International Conference on Electrical Machines; ICEM 2010. Rome: IEEE; 2010

[98] Prankl H, Nadlinger M, Demmelmayr F, Schrödl M, Colle T, Kalteis G. Multifunctional PTO generator for mobile electric power supply of agricultural machinery. In: International Conference on Agricultural Engineering; 2011. Hannover; VDI Berichte 2124: 2011

[99] Wuebbels R. Machine for harvesting stalk-like plants with an electrically driven cutting mechanism. US Patent 0174552 A1; 2012

[100] Stoss KD, Sobotzik J, Shi B, Kreis ER. Tractor power for implement operation – mechanical, hydraulic and electrical: an overview. In: Agricultural Equipment Technology Conference ASABE Distinguished Lectures Series **37**; 28-30 January, 2013. 2013. pp. 1-25; ASABE Publ. no. 913C0113

[101] Laguens M. Potential for energy savings through hybridization of agricultural tractors. Engineering Degree Dissertation, Madrid: Tech. Univ.; 2014

[102] Zhitkova S, Felden M, Franck D, Hameyer K. Design of an electrical motor with wide speed range for the in-wheel drive in a heavy duty off-road vehicle. In: International Conference Electrical Machines (ICEM); 2-5 September; Berlin, Germany; 2014. pp. 1076-1082

[103] Hahn K. High voltage electric tractor-implement interface. SAE International Journal of Commercial Vehicles. 2008;**1**:383-391

[104] Keil R, Shi B, Sobotzik J. JD 6210RE-tractor/implement electrification and automation. Antriebsysteme 2013-Elektrik, Mechanik und Hydraulik in der Anwendung. VDE Verlag; 2013

[105] Buning EA. Electric drives in agricultural machinery—Approach from the tractor side. In: Key Note Report. 21st Annual Meeting of the Club of Bologna;November 13-14. Bologna, EIMA International; 2010

[106] Somà A, Bosso N, Merlo A. Electrohydraulic hybrid lifting vehicle. Patent, WO2011128772; 2011

[107] Lin T, Wang Q, Hu B, et al. Development of hybrid powered hydraulic construction machinery. Automation in Construction. 2010;**19**:11-19

[108] Wang J, Yang Z, Liu S, Zhang Q, Han Y. A comprehensive overview of hybrid construction machinery. Advances in Mechanical Engineering. 2016;**8**(3):1-15

[109] Moreda GP, Muñoz-García MA, Barreiro P. High voltage electrification of tractor and agricultural machinery—A review. Energy Conversion and Management. 2016; **115**(2016):117-131

[110] Zou NW, Dai QL, Jia YH, et al. Modeling and simulation research of coaxial parallel hybrid loader. Applied Mechanics and Materials. 2010;**29**:1634-1639

[111] Wang F, Zulkefli MAM, Sun Z, et al. Investigation on the energy management strategy for hydraulic hybrid wheel loaders. In: ASME 2013 Dynamic Systems and Control Conference; 21-23 October 2013; Palo Alto, CA.V001T11A005 (10 pp.). New York: ASME;2013

[112] Sun H, Jiang JQ. Research on the system configuration and energy control strategy for parallel hydraulic hybrid loader. Automation in Construction. 2010;**19**:213-220

[113] Ochiai M, Ryu S. Hybrid in construction machinery. Paper no. 7-1. In: Proceedings of the 7th JFPS International Symposiumon Fluid Power; 15-18 September 2008. Toyama, Japan; Tokyo: JFPS: 2008. pp.41-44

[114] Ochiai M. Technical trend and problem in construction machinery. Construction Machinery. 2002;**38**:20-24

[115] Riyuu S, Tamura M, Ochiai M. Hybrid construction machine. Patent 2003328397, Japan; 2003

[116] Grammatico S, Balluchi A, Cosoli E. A series-parallel hybrid electric powertrain for industrial vehicles. In: IEEE Vehicle Power and Propulsion Conference (VPPC);1-3 September 2010. Lille; 2010

[117] Sierks-Schilling B. 12MTX Hybrid—A major project for the environment. 2009. http://www.building-construction-machinery.net/

[118] Zhou N, Zhang E, Wang Q, et al. Compound hybrid wheel loader. Patent CN102653228A, China; 2012

[119] Flint J. A different kind of hybrid from John Deere. October 2013. FarmProgress.com

[120] Anderson E, John Deere 644K Hybrid Drivetrain Overview, Performance, & Developmental Analysis. In SAE 2013 Heavy Duty Vehicles Symposium Technologies for High Efficiency & Fuel Economy; Rosemont, IL; September 2013

[121] Rebholz W. New hydrostatic/mechanical power-split CVT for use in construction machinery. Symposium on Gearbox in Vehicle; 2010

[122] Bohler F, Thiebes P, Geimer M, Santoire J, Zahoransky R. Hybrid system for industrial application. SAE-NA Paper Series; 2009

[123] Kagoshima M, Sora T, Komiyama M. Hybrid construction equipment power control apparatus. Patent7069673, USA; 2006

[124] Kagoshima M. The development of an 8 tonne class hybrid hydraulic excavator SK80H. Kobelco Technology Review. 2012;**31**:6-11

[125] Kagoshima M, Komiyama M, Nanjo T, Tsutsui A. Development of new hybrid excavator. Kobelco Technology Review. 2007;**27**:39-42

[126] Kwon TS, Lee SW, Sul SK, et al. Power control algorithm for hybrid excavator with supercapacitor. IEEE Transactions on Industry Applications. 2010;**46**:1447-1455

[127] Edamura M, Ishida S, Imura S, Izumi S. Adoption of electrification and hybrid drive for more energy-efficient construction machinery. Hitachi Review. 2013;**62**(2):118-122

[128] Profile Komatsu Corporate (KC). PC200-8 hybrid hydraulic excavator contributes to reducing CO_2 emissions. Views. 2008;**3**:4-5

[129] Cho S, Yoo S, Park C. Development of a mid-size compound type hybrid electric excavator. EVS27 Barcelona, Spain, November 17-20, 2013

Chapter 2

Development of Bus Drive Technology towards Zero Emissions: A Review

Julius Partridge, Wei Wu and Richard Bucknall

Additional information is available at the end of the chapter

http://dx.doi.org/10.5772/68139

Abstract

This chapter aims to provide a comprehensive review of the latest low emission propulsion vehicles, particularly for bus applications. The challenges for city bus applications and the necessity for low emission technologies for public transportation are addressed. The review will be focusing on the London bus environment, which represents one of the busiest bus networks in the world. The low emission bus applications will be analysed from three main areas: hybrid electric buses, battery electric buses and fuel cell buses. This summarises the main technologies utilised for low emissions urban transportation applications. A comprehensive review of these low emission technologies provides the reader with a general background of the developments in the bus industry and the technologies utilised to improve the performance in terms of both efficiency and emission reduction. This will conclude with a summary of the advantages and disadvantages of the three main technologies and explore the potential opportunity of each.

Keywords: low emission drive, battery bus, hybrid electric bus, fuel cell bus, vehicle performances

1. Introduction

Over the past 100 years, the bus industry has come to be dominated by diesel powered buses due to their increasingly low cost and greater maturity of the technology. However, this comes at an environmental cost, for example, over 600 kt of CO_2 was emitted by London's bus fleet in 2015 [1]. It is these carbon emissions and their link to climate change that have provided one of the major drivers in recent years to develop and deploy alternative technologies for bus propulsion [2]. Other emissions associated with diesel vehicles such as NO_x and particulates

INTECH
open science | open minds

© 2017 The Author(s). Licensee InTech. This chapter is distributed under the terms of the Creative Commons Attribution License (http://creativecommons.org/licenses/by/3.0), which permits unrestricted use, distribution, and reproduction in any medium, provided the original work is properly cited.

have provided a local driver to change due to their detrimental impacts on human health [3–5]. In 2008, it was estimated over 4000 deaths were brought forward as a result of long-term exposure to particulates in London [6]. In order to combat these concerns, many cities have introduced measures such as the 'low emission zone' in London and emission control targets [7]. London is to introduce the first ultra-low emissions zone (ULEZ) in 2020, which, amongst other targets will aim to replace conventional diesel powered buses with low emissions alternatives [8, 9]. Despite this drive for change, it is evident that finding a replacement for diesel buses is not simple. In addition to the low cost, simplicity, reliability and maturity of the technology, diesels also offer excellent characteristics to meet the required power demands and operational needs of city buses. It can be seen from **Figure 1**, the diesel engine that is a type of internal combustion engine (ICE) provides high output powers and uses energy dense fuel making them ideal for both the range and operating times expected of city buses and also for meeting the high transient power requirements during acceleration.

In order to address the environmental concerns posed by diesel buses, a number of technologies are being investigated and implemented. The most widespread of these are diesel-hybrid buses, which make use of an on-board energy storage system to effectively recycle captured kinetic energy obtained through regenerative braking. Although hybrid buses are capable of significantly reducing fuel consumption, they are still reliant on diesel as the primary fuel source and hence do not address the fundamental problems associated with emission that come from using diesel as a fuel. As such, there has in recent years been an increased focus on the development of zero emissions buses, with two main competing technologies. These are battery electric buses and hydrogen fuel cell (FC), both of which exhibit zero operating emissions, hence eliminates the environmental and health issues associated with diesel buses [11]. Such technology solutions are less mature and result in significant changes to the

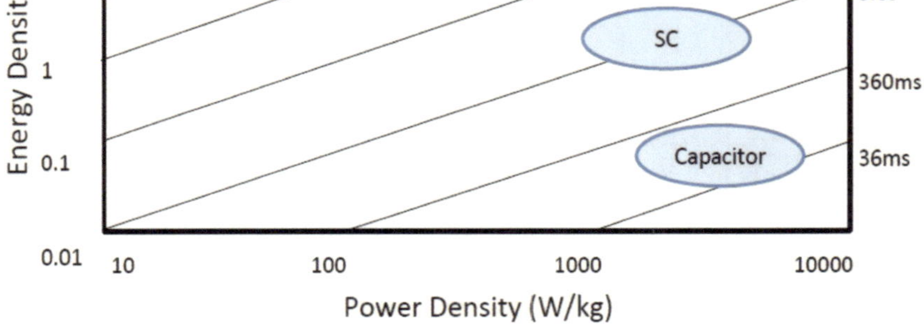

Figure 1. Comparison of various technologies for the power and energy densities (based on Ref. [10]).

propulsion system. Although these technologies have been deployed in operational bus fleets, there remain a number of barriers to widespread deployment.

London has one of the most comprehensive and busiest public transport networks in the world, operated by Transport for London (TfL). There are over 9000 buses in operation [12], which are estimated to account for 21% of the CO_2 emissions in London [7], 63% of NOx and 52% of PM_{10} particulate emissions [13]. It is reported that the TfL bus fleet carries 6 million passengers each working day, which the number of bus passenger journeys grew by 64% between 2000 and 2013 and is continuing to increase [14]. The Greater London Authority (GLA) has introduced a number of strategies in an attempt to reduce emissions from buses, part of which is the London hybrid bus project which aims to replace the conventional bus fleet with diesel hybrid buses [7, 15]. This is to be furthered with the introduction of the ultra-low emissions zone (ULEZ) in 2020, which, amongst other targets will require all 3000 double-decker buses operating in the ULEZ to be diesel hybrid and all 300 single decker buses to be zero emissions [8, 9, 16]. Since 2004, a number of technologies have been deployed as part of the operational bus fleet, as shown in **Figure 2**, as a means of reducing emissions. London has been used as a case study throughout this chapter due to both the comprehensive bus network and the operational deployment of new technologies.

Within this chapter, the development of low emission bus propulsion technologies will be discussed, through the evolution of diesel to diesel hybrid buses and onto the development and deployment of battery electric and FC buses. The aim is to outline the benefits of such technologies and the barriers that exist to their widespread implementation from both a technical and economic perspective. Part 2 discusses the implementation of diesel electric hybrid buses

Figure 2. Timeline of the milestones for the London low emission bus deployment.

and their evolution from diesel buses. Parts 3 and 4 consider battery electric buses and fuel cell buses, respectively, whilst part 5 provides a comparison of these emerging technologies.

2. Diesel hybrid bus

2.1. Basic principles of diesel electric hybrid buses

The principle difference between diesel hybrid buses and diesel buses is the inclusion of an electrical energy storage system in conjunction with an electrical motor/generator. The primary source of energy is still the diesel engine; however, the inclusion of the electrical system provides a number of advantages such as facilitating regenerative braking and allowing reduced idling time [17]. The utilisation of a hybrid system results in improvements fuel efficiency and emissions, although these come at the price of additional cost and complexity [17].

The integration of the electrical energy system can be utilised through a number of configurations, with the common options being the series, parallel and series-parallel hybrid configurations, as shown in **Figure 3**. In a series hybrid drivetrain, the mechanical output from the diesel engine is converted into electrical power via a generator when operating at its most efficient loading. This is supplemented with a battery to provide for the electric drive motor

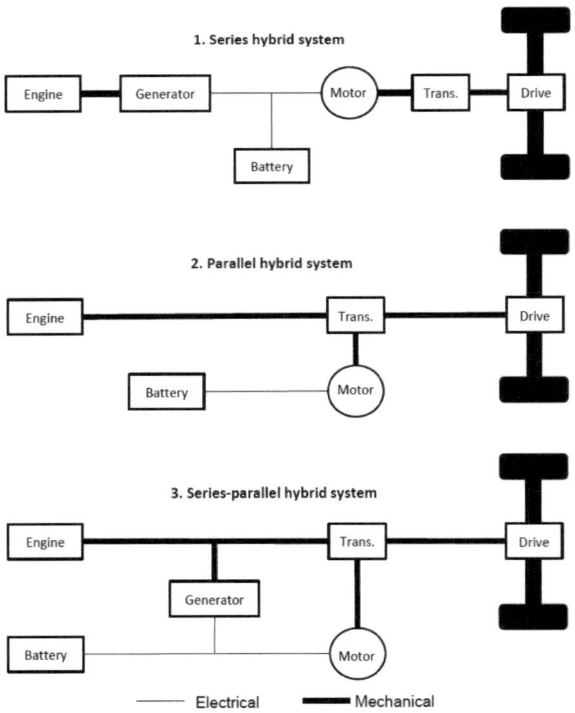

Figure 3. Schematic of the three available layouts for the propulsion system of a diesel/electric hybrid drive train.

requirements. Since the propulsion needs are met by an electric motor, this results in the complete decoupling between the diesel engine and the wheels, meaning that engine control is not dependent on vehicle speed so offering additional flexibility [18]. This is a major advantage of series hybrid drivetrains, where the engine can operate at any point on its speed-torque map, which is impossible for conventional vehicles. Therefore, the engine is capable of constantly operating at near optimum load, which minimises fuel consumption and emission [19].

The parallel hybrid configuration maintains the direct mechanical link between the diesel engine and the wheels, using the battery for regenerative braking and supplementing the peak power demands. The main advantages over the series hybrid are that the additional generator is no longer needed so has higher efficiency as well as reducing the size of the required drive motor. The parallel configuration, however, does not decouple the diesel engine from the wheels and hence operation is directly linked to the vehicle speed hence for low speed city operation the ICE will often operate at a low efficiency [20]. As a result, the parallel configuration is more appropriate for longer distance and higher speed routes. The series-parallel hybrid can operate in either the series or parallel configurations and so can utilise the advantages of both systems; however, the additional complexity and capital cost of the system mean that they are currently not a viable option for transportation applications [19]. The most popular option for city buses is the series configuration due to the simplicity of a single drive system as well as higher efficiency during city driving where buses have a start-stop traffic pattern with generally low speed operation [19].

The benefits offered by the hybridisation of the drive system relate to the increase in fuel economy and reduction in emissions compared to a diesel bus and can be attributed to the following points.

- On average buses idles for around 30–44% of urban driving time [21]. By using a hybrid system, the vehicle can turn off the engine to prevent idling and low loads because it can use the electrical energy storage and motor for initial acceleration. This can save 5–8% of fuel consumption [17].

- A significant amount of energy is lost and dissipated by heat due to friction during conventional braking. When a hybrid vehicle is braking, the drive motor can work as generator to charge the electrical energy storage system and thus recycle some of the energy used to propel the bus. Typically, 10–20% of the kinetic energy is recovered.

- In a conventional bus, the diesel engine needs to be large enough to provide for all of the peak transient power demands. A hybrid vehicle is able to use the electrical system to provide for a portion of these peak demands, and therefore, the engine can be downsized [17, 19].

- A diesel engine operates at its lowest efficiency during low load and low speed operation. The electrical system can drive the electric motor to power the bus during low load and start-up to avoid this. It is expected that diesel hybrid technology can achieve reductions of between 24 and 37% CO_2 emission [22], 21% to NO_x emission and 10% to fuel consumption compared with conventional diesel buses [7, 15].

In contrast to these benefits, the hybridisation of the drive system has a number of drawbacks. These predominately amount to the additional capital cost, where a diesel hybrid typically costs £300,000, this is £110,000 more than a conventional diesel bus and constitutes an increase of about 50% [23]. The additional complexity of both the drive system and its control results in additional maintenance time and cost, where a diesel hybrid typically requires 50% more maintenance time than a conventional diesel bus [22].

2.2. Case study 1: TFL

Initially a trial consisting of eight diesel hybrid buses was carried out in 2007 and was found to have very high (96%) customer support [24]. After analysing the trial, the official deployment of diesel hybrid buses began in central London. The number of diesel hybrid buses has steadily increased, where in 2015, more than 1200 diesel hybrid buses were in operation in London, as can be seen in **Figure 4**, and exceed the target of 1700 in 2016 [12]. This consists of old buses redesigned for hybrid operation and new designs such as the new Routemaster.

The impact of the deployment of the low emission bus fleets has already begun to have an impact on emissions in London, as shown in **Figure 5**. In the last few years, emissions of NO_x and CO_2 have begun to drop due to the introduction of diesel hybrid buses into the TfL fleet and the retrofitting of selective catalytic reduction measures into the existing fleet. The level of PM100 emissions dropped considerably due to the introduction of PM filters in the early 2000s. It is expected that these will continue to drop as further deployment of diesel hybrid and zero emissions vehicles continues.

The performance of the diesel hybrid bus fleet in London is very variable, as might be expected due to differing models and routes. It was claimed that the average Euro V bus

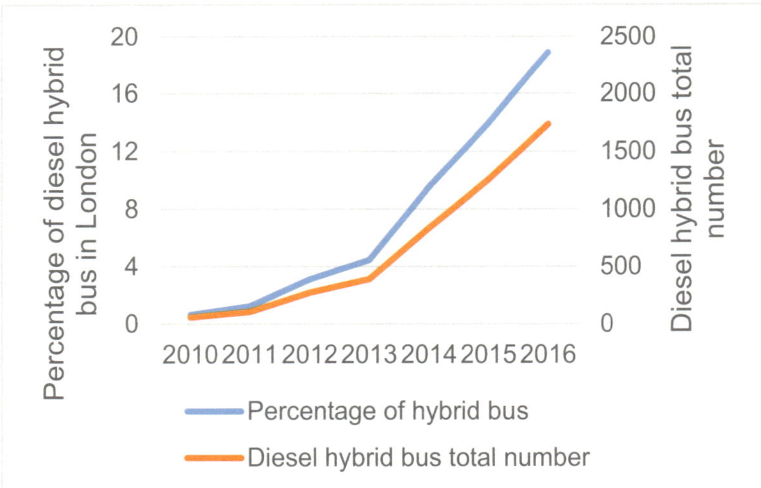

Figure 4. Total number and percentage of the TfL bus fleet of diesel-hybrid buses in operation. Data from Ref. [12].

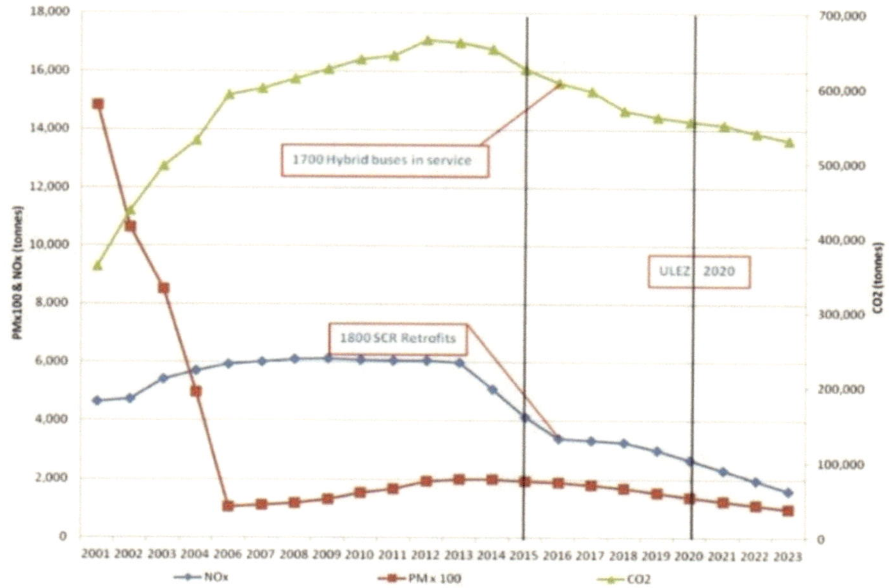

Figure 5. Expected reduction in CO_2 and NO*x* emissions from the TFL bus fleet with the deployment of diesel/electric hybrid buses [1, 15, 25].

achieved a fuel economy of 32.9 l/100 km in London [9]. The reported fuel economy of diesel hybrid buses operating in London is presented in **Table 1**. As may be expected, the type of bus and bus route significantly affects the fuel economy, where a single decker bus generally exhibits a higher fuel economy than a double-decker bus. It was found that the introduction of diesel hybrid technology improved the fuel economy on nearly all routes; however, there were a couple of discrepancies to this, such as on the E8 bus route where the fuel economy actually decreased. The introduction of the new Routemaster bus appears to provide a slight improvement over previous diesel hybrid buses; however, there appears to be significant discrepancies between the recorded and expected performance. Results released by TfL in 2014 suggest a fuel economy in the range of 38.2–45.6 l/100 km, whereas it is claimed by the manufacturer that a fuel economy of 24.4 l/100 km was recorded on the 159 bus route. Unfortunately, the details for these results are not available and so it is difficult to determine the validity of the results. This discrepancy could be the result of a number of factors such as the route topology, traffic conditions, driving style and passenger conditions.

In summary, TfL has successfully introduced a large number of diesel hybrid buses into their bus fleet. This has resulted in a decrease in the emissions associated with the bus fleet, with considerable further reductions expected. It provides an example of the successful deployment of diesel hybrid buses into a large operational bus fleet to achieve reductions in emissions and fuel consumption. However, the increased cost and system complexity remain problematic.

Bus type	Route	Diesel	Diesel hybrid	Year	References
		Fuel economy (l/100 km)	Fuel economy (l/100 km)		
Single decker (Euro V)	276	44.8	43.5	2010	[26]
	360	36.7	34.9		
	371	34.1	26.7		
	E8	35.3	42.2		
	129	47.1	33.6		
Double decker (Euro V)	141	60.1	50.4	2010	[26]
	328	65.7	54.3		
	24	49.6	42.2		
	482	50.4	34.9		
	16	50.4	39.2		
New routemaster (Euro V)	11	60.1	38.2	2014	[27]
	24/390	52.3	38.2		
	9	72.4	45.6		
	148	56.5	40.9		
	10	64.2	43.5		
	159	Not available	24.4	2013	[28]

Table 1. Available data for diesel hybrid bus fuel economy in London. The values for l/100 km have been converted from miles per gallon using $Litres_{100\ km} = (100*4.54609)/(1.609344*mpg_{uk})$.

3. Battery electric bus

3.1. Overview of electric buses technology

The battery electric bus, often described as a pure electric bus, uses an electric motor for propulsion and a battery for energy storage [29]. In most cases the battery is the primary energy source, although for trolley buses power is delivered from overhead cables during operation.

The configuration for electric buses is typically fairly straightforward since it is basically a battery driving an electric motor to propel the vehicle [30], as shown in **Figure 6**. During braking it is also possible to make use of regenerative braking to recharge the battery during braking. The main battery technologies that have been used in transportation are Ni-MH, Zebra ($Na\text{-}NiCL_2$) and lithium batteries [31]. The most promising of these are the lithium batteries, where three main categories exist, these being Li-ion, lithium polymer (LiPo) and Lithium-iron-phosphate ($LiFePO_4$) batteries [32]. Most current buses use lithium-based batteries [33] due to their high power and energy densities and fast charging capabilities, although their high cost is still problematic [32]. A problem faced by all battery technologies is their cycle life; typically, these are short

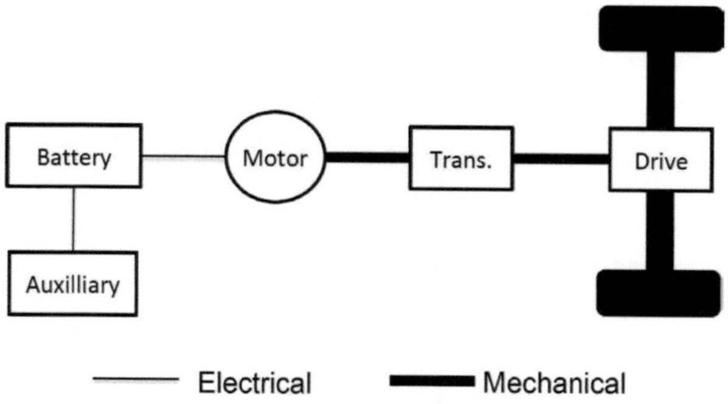

Figure 6. Battery electric drive bus basic configuration.

and hence require relatively regular replacement [34]. In addition to a battery pack, some buses utilise supercapacitors in conjunction with a battery as supercapacitors are much more effective in shielding batteries from high current load and thus increase battery life [35]; however, their low energy density means they are unsuitable to be used as the primary energy source, as shown in **Figure 1**. They do, however, have several key advantages over existing battery technologies, such as very high power densities and discharge rates as well as very long cycle life [34]. There is no simple answer to which battery technology is best, as it will depend on the application. Mahmoud et al. [36] carried out a detailed comparison study of different electric powertrains and concluded that a single technological choice would not satisfy the varied operational demands of transit services because electric buses are highly sensitive to the energy profile and operational demands. Electric buses are zero emission at the point of use and therefore offer great emission savings particularly in terms of local air pollution when compared to ICE or hybrid buses, as well as very high efficiency. However, there are a number of barriers to widespread deployment, the main ones are recharging time, vehicle range, infrastructure and cost [34].

Battery electric buses normally operate in one of two different forms: opportunity and overnight [32]. Opportunity e-buses have a smaller energy storage capacity that offers limited range but can be charged much quicker (5–10 minutes); while overnight e-buses have a much larger energy storage but at the cost of longer charging time (2–4 hour) [36]. These represent two different approaches for electric buses in the urban environment. The opportunity approach aims to minimise the weight of the battery pack by utilising frequent and fast recharging at points along the bus route, such as bus stops or the end of route [32]. This holds the promise of high efficiency and lower projected bus costs but requires a comprehensive recharging network [37]. Route flexibility of the bus is, however, limited, as it is required to follow the assigned bus route to recharge the battery. The overnight method utilises a large energy storage system to extend the range so that the bus can drive the entire route/day without recharging [37]. This holds the promise of greater route flexibility and convenience as well as utilising a centralised recharging infrastructure, but suffers from passenger loss due to increased battery weight as well as battery lifetime issues [38] and battery cost [34]. An alternative approach is offered by the Trolleybus, which has a small battery but receives power from overhead cables along the assigned route. This overcomes problems associated with range and recharging times but is very limited in terms of route flexibility.

The process of recharging a battery electric bus can be completed through plug-in (conductive), wireless (inductive) or catenary (overhead power lines) charging. Plug-in charging requires a direct connection through a power cord [39] and is well-suited to overnight bus charging, but can be used in some instances for opportunity charging. This is popular due to its simplicity and high efficiency [39]. Wireless charging relies on induction between two coils, this is better suited to opportunity buses where recharging can take place along the route without the need for a physical connection [39], such as the PRIMOVE bus where charging is carried out at each end of the route and at five intermediate stops [40]. This form of charging, however, suffers from increased charging times and relatively low efficiency [39]. The trolleybus uses overhead catenary to provide power to the bus [41]. This type of charging exhibits high efficiency but requires an extensive network of overhead cables.

Table 2 shows a selection of operating pure electric buses in different locations and utilise a number of battery technologies and operating approaches. In 2015, there were an estimated 150,000 battery electric buses, mostly located in China, with a sixfold increase between 2014 and 2015 [42]. The electric bus market is growing quickly where it had a 6% share of global bus purchases in 2012 but is forecasted to grow to 15% by 2020 [43]. Battery electric bus development has been carried out all over the world with the largest shares in China, Europe and North America [44]. It is clear that some of the buses listed in **Table 2** utilise more than one mode of operation to provide for the operational power requirements, such as the complete coach works bus, which uses both overnight and opportunity charging. The differences in

Manufacturer	Length	Capacity	Battery type	Battery capacity	Type, range	Deployment location
ABB TOSA	18 m	135	Lithium titanate oxide	38 kWh	Trolley, on-route	Switzerland
BYD	12 m	40	BYD Iron Phosphate	324 kWh	Overnight, 250 km	Worldwide
Complete Coach Works	12 m	37	Lithium-iron Phosphate	213 kWh	Overnight/ opportunity, 145 km	US
EBusco	12 m	76	Lithium-iron Phosphate	242 kWh	Overnight, 250 km	China, Finland
Hengtong EBus	12 m	70	Lithium Titanate	60.9 kWh	Opportunity, 39 km	China
New Flyer	12 m	40	Lithium-Ion	120 kWh	Opportunity, 72 km	US, Canada
Primove	12 m	44	Lithium-Ion	60 kWh	Wireless, on-route	Germany
Proterra	10 m	35	Lithium Titanate	74 kWh	Opportunity, 42 km	US
Siemens	8 m	40	Lithium-iron Phosphate	96 kWh	Trolley, on-route	Austria
Sinautec	12 m	41	Ultra-Cap and Battery	5.9 kWh	Trolley, on-route	China

Table 2. Selection of operating electric bus models worldwide [40].

operating regimes are reflected in the sizing of the batteries and as a result the range of the buses, where they vary from 5.9 kWh for the trolleybus design to >300 kWh for overnight charging. This will have a significant impact in terms of the bus's battery costs; however, the charging infrastructure for overnight charging does not need to be as comprehensive as for the alternative methods.

3.2. Case study: London electric buses

London has been working on overnight e-bus demonstrations since 2012 and is also investigating the potential of opportunity e-bus technologies. From the overnight e-bus perspective, TfL has collaborated with BYD, which is one of the largest electric bus manufacturer in China, to test the potential of battery electric buses in London, starting from 2012 [45]. The first two battery electric buses were handed over to TfL in 2013 and then entered daily service on two central London routes, numbers 507 and 521, which were the first battery electric buses in London. These single-decker 12-metre BYD buses utilise Lithium-Iron-phosphate batteries and have demonstrated a range in excess of 250 km on a single charge in real world urban driving conditions [46]. The 507 and 521 bus routes are relatively short commuter service routes and were chosen so that the bus can start operating in the morning peak alongside the diesel bus fleet and return to the depot to recharge during the day before resuming service for the evening peak [34, 47]. The battery takes 4–5 hours to recharge when fully discharged and has been designed for a cycle life of more than 4000 cycles, meaning a 10-year battery lifetime under normal operating conditions [48]. The trail fleet was extended to six buses in the summer of 2014. The trail buses in London not only provide a zero emission environmental benefit but also have shown promising result in terms of both technical and economic performance, and hence TfL has taken further steps towards adopting this new clean technology in the capital. The development timeline and future plans for London electric buses are plotted in **Figure 7**.

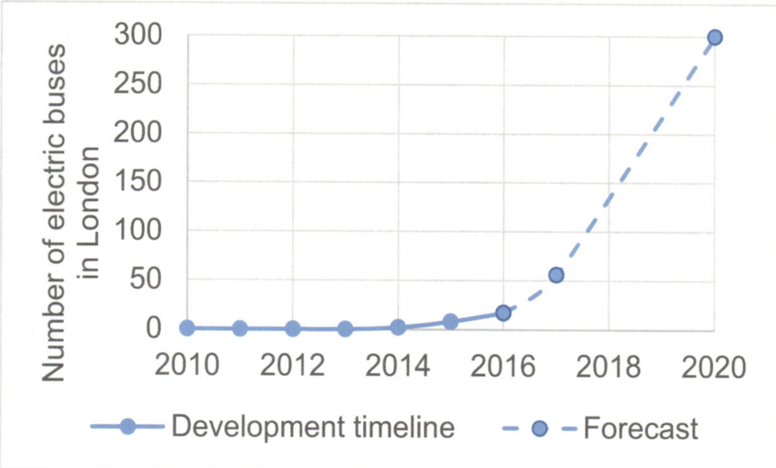

Figure 7. Number of electric buses in London.

The latest data in 2016 showed that there are currently 22 battery electric buses operating in London including 17 single-decker battery electric buses and five double-decker battery electric buses. This is a world first for double-decker battery electric buses, as shown in **Figure 8**, and entered service in May 2016. These are 10.2 m buses with a capacity of 81 passengers and a claimed range of 303 km. The battery is a Lithium-Iron-Phosphate battery with a capacity of 320 kWh [49]. They utilise a combination of both overnight and opportunity e-bus technology and will operate on route 69 in Central London. They will use a high powered wireless inductive charging system to top up their battery system at the beginning and end of this route to keep the bus operating throughout the entire day [50]. The recent double-decker electric buses have used wireless charging technology as part of innovative charging technology development. However, this is still far from a mature technology and requires a massive recharging infrastructure network [51]. The electric buses in London have shown promising performance on short commuter routes; however, pure e-buses are still best suited for shorter routes with operational flexibility and scope to recharge them in inter-peak periods due to the limit of present battery capacity and recharging technology [52].

In 2015, BYD and Alexander Dennis (ADL) announced a partnership to provide 51 further single-decker buses to route operator Go-Ahead with an expected delivery in late 2016 [53]. BYD will provide the batteries and electric chassis technology, and ADL will provide the bus body-building technology [54]. The cost of each bus is expected to be £350,000 [55].

In summary, the recent development and deployment of battery electric buses in London have shown that electric buses are technically feasible. It can be seen that electric buses will also have an important role to play in the coming ULEZ implementation in 2020. However, more time is needed to evaluate the actual performance and address the key challenges facing electric buses such as limitations of battery technology that restricts range.

Figure 8. The first electric double-decker bus in the world (photo from Business Green, 2016).

http://dx.doi.org/10.5772/68139

4. Hydrogen fuel cell hybrid bus

4.1. Basic theory

Hydrogen fuel cells (FCs) are considered a clean energy source with the main benefits over ICEs of zero harmful emissions during operation and high efficiency [56]. Although many types of FCs exist, this paper will only consider the application of FCs in transportation, considering the operating temperature, start-up time and technology maturity, Proton Exchange Membrane Fuel Cell (PEMFC) offer most promising solution [57]. Significant research into solid oxide fuel cells (SOFCs) in transportation has been carried out [58–60], although these have yet to been applied in real world bus applications. A PEM FC uses hydrogen as the fuel, which, through an electrochemical reaction with oxygen (usually from air) generates electricity with water as the only by-product from the chemical process [61]. By replacing the internal combustion engine in conventional buses, FCs can be used as the primary energy source to power a bus with electrical energy, therefore, achieving zero operating emissions. An additional advantage over ICE's comes from the higher efficiencies exhibited by FCs [62, 63]. However, there are a number of barriers that need to be overcome before widespread deployment can be achieved. These are primarily cost and infrastructure [64, 65]. FC powered buses cost approximately five times more than a conventional diesel bus with the similar power output [66], where they typically cost in excess of £1,000,000 [67], due primarily to the expensive FC stack and the small scale of production [68]. In addition, the widespread deployment of FC buses would require a significant investment in hydrogen refuelling infrastructure [64]. The implementation of FC buses has shown that the technology is a promising solution for zero emissions buses if these barriers can be overcome.

Figure 9 shows the configuration usually used in FC vehicles. The basic drive train utilises a FC to power the propulsion motor; however, FCs are not well suited to providing for the transient power demands associated with city driving buses [69–73]. As such, most FC buses utilise a form of energy storage in a series configuration to both address this and also to

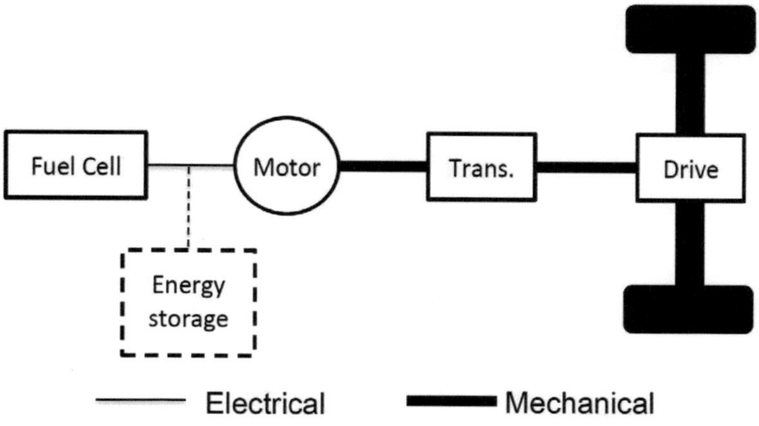

Figure 9. Simplified architectures of FC drivetrain.

utilise regenerative braking [74]. An additional benefit of such an approach is that the size of the expensive fuel cell stack can be reduced [75]. The energy storage implemented is usually either electrochemical battery technology such as Li-ion or NiCd batteries or electrostatic supercapacitors (sometimes referred to as ultracapacitors). The choice between these depends on the particular design and requirements of the system, with batteries offering reasonable power and energy densities although they have a relatively short cycle life and supercapacitors offering poor energy densities but excellent power densities, as shown in **Figure 1**. Additionally, supercapacitors have very long lifetimes of up to 40 years [31].

In a series configuration, there are three main modes of operation that can be utilised to provide for the buses power demands, as shown in **Figure 10**. Although these are the main modes of operation, the way these modes are utilised will depend on the control strategy implemented [76].

- Mode 1: The SC discharges to supplement the FC to provide for high transient power demands. This type of operation is expected to occur under heavy loads such as during acceleration or going uphill.
- Mode 2: The FC will both power the load and use excess power to charge the SC. This is expected to occur under low loads, when the FC power output is higher than the required load.
- Mode 3: The power from the FC and generated power from regenerative brake will both be used to charge the SC. This is only expected to occur when the drive motor is operating as a generator in the regenerative brake mode.

There have been a number of projects aimed at utilising FC technology for bus propulsion applications. **Table 3** lists many of the projects currently in operation along with the FC size and energy storage used. The projects are split into two main categories depending on the relative size of the FC and energy storage systems. The majority of the current projects are FC

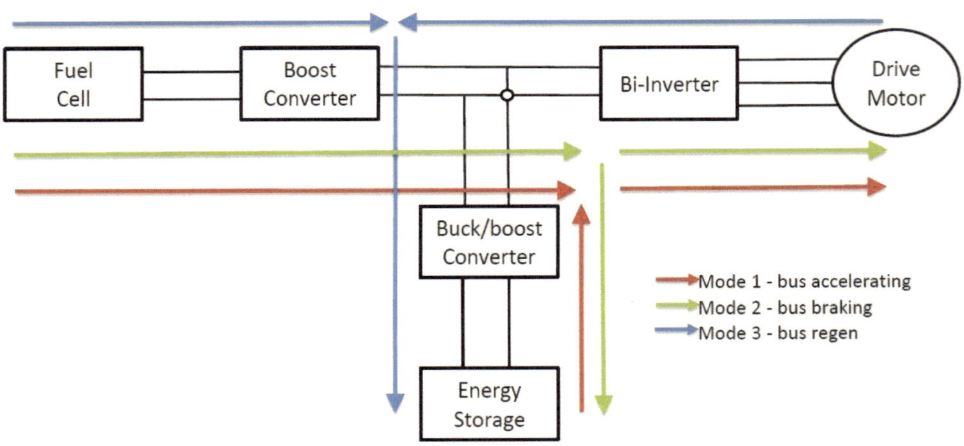

Figure 10. Modes of operation for a series hybrid FC drive train [77].

dominant, whereby the FC is expected to provide for the majority of the propulsion needs. Alternatively there a few examples of battery dominant hybrids, where the battery is the main source of power with the FC used as a supplementary power source. It was announced in 2017 that the JIVE project is to implement 142 buses across nine European cities with 56 new FC buses in the UK, which will be the first large scale validation project of FC bus fleets [78].

4.2. Case study: TfL FC bus on the RV1 bus route

London has been involved with the testing and deployment of FC buses, **Figure 11** shows the evolution of FC bus implementation in London. Initially, this was through the EU funded Clean Urban Transport for Europe (CUTE) project, which aimed at introducing hydrogen FC buses into European cities, where a test run of three buses were operated on the RV1 bus route between 2004 and 2006, this was increased to five buses from 2007 to 2009 [83]. London is now part of Clean Hydrogen in European Cities (CHICs) project with the first deployment in full service of the next generation of FC bus in 2011 and is expected to continue until 2019. There are currently eight Hydrogen buses operating in Central London as part of the CHIC project, fully covering the RV1 bus route, which is 9.7 km in length [83]. It is expected that by 2017 a further two buses developed as part of the 3Emotion project will be deployed through Van Hool [84]. The buses operate for 16–18 hours/day, before returning to the depot for refuelling at the central depot, which takes <10 minutes [85]. The workshop, which is responsible for routine maintenances and hydrogen management, was specifically designed and built for hydrogen

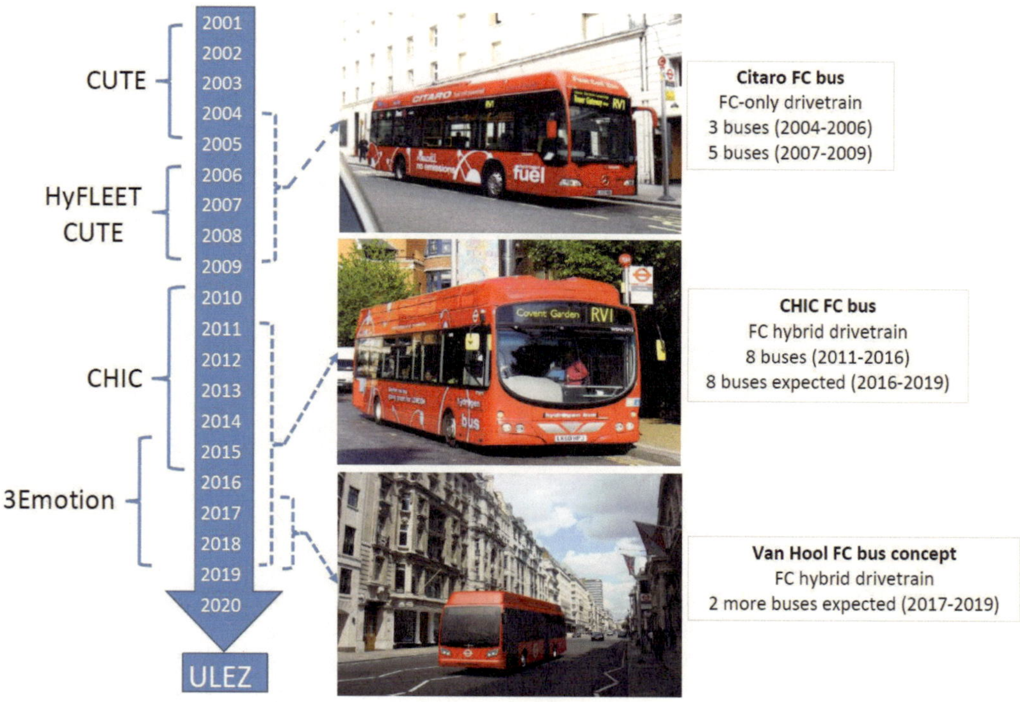

Figure 11. London FC hydrogen bus development timeline (bus photos from Citaro, TfL, Van Hool, 2016).

Project	Fleet	Year	Location	Length (m)	FC size (kW)	Battery type	Battery size (kWh)	Drive type
JHFC	2	2006	Tokoname, Japan	10.5	180	Nickel Metal Hydride	Not available	FC dominant hybrid
University of Delaware	2	2007	Dewark, US	6.7	40	NiCad	60	Battery dominant hybrid
TriHyBu	1	2009	Neratovice, Czech Republic	12	48	Lithium Ion	26	Battery dominant hybrid
BurbankBus	1	2010	Burbank, US	10.7	32	Lithium Titanate	54	Battery dominant hybrid
HySUT	2	2010	Tokyo, Japan	10.5	180	Nickel Metal Hydride	Not available	FC dominant hybrid
NFCBP	1	2010	San Francisco, US	12.2	32	Lithium Ion	Not available[3]	FC APU Compound
Toyota FCHV	1	2010	Toyota City, Japan	10.5	180	Nickel Metal Hydride	Not available	FC dominant hybrid
NFCBP	4	2010	Hartford, US	12.2	120	Lithium Ion	17.4[3]	FC dominant hybrid
CHIC	8	2010	London, UK	12	75	Supercapacitor	0.5	FC dominant hybrid
CHIC	3	2011	Milan, Italy	11.9	120	Lithium Ion	26	FC dominant hybrid
SunLine[1]	6	2011	Thousand Palms, US	12.2	150	Nanophosphate Li-ion	11	FC dominant hybrid
NFCBP	12	2011	Multi-city, US	12.2	120	Lithium Ion	17.4	FC dominant hybrid
CHIC	4	2011	Cologne, Germany	18.4	150	NiMeH and Supercapacitor	23 and 0.6	FC dominant hybrid
CHIC	5	2011	Aargau, Switzerland	11.9	120	Lithium Ion[4]	26.9[4]	FC dominant hybrid
CHIC	5	2012	Oslo, Norway	13	150	Lithium Ion	17.5	FC dominant hybrid
NIP, CHIC	6	2012	Hamburg, Germany	12	120	Lithium Ion	26	FC dominant hybrid
CHIC	5	2013	Bolzano, Italy	11.9	120	Lithium Ion	26	FC dominant hybrid
HyTransit, HighVLO City	14	2014	Aberdeen, UK	12.2	150	Not available	Not available	FC dominant hybrid
HighVLO City	5	2014	Brussels, Belgium	12.2	150	Not available	Not available	FC dominant hybrid
NFCBP[2]	1	2014	Austin, US	10.7	30	Lithium Titanate	54	Battery dominant hybrid
NFCBP[2]	1	2014	Birmingham, US	9.8	75	Lithium Titanate	54	Battery dominant hybrid

Notes: [1][79], [2][80], [3][81], [4][82].

Table 3. All active fuel cell bus demonstration project in 2016.

FC buses [86]. The hydrogen has been transported in liquid form to the depot and converted into gaseous form to refuel buses [83], it is then stored on site in gaseous form at 500 bar [86].

The buses themselves have developed throughout this project, where the first generation was powered only by a FC. These utilised a 250 kW fuel cell [82] and achieved a hydrogen economy of 18.4–29.1 kg H_2/100 km [87]. The buses deployed as part of the CHIC project utilised a series hybrid configuration, with a 75 kW PEM FC from Ballard and a 0.5 kWh Bluways supercapacitor energy storage system [88]. This introduction of the hybrid system significantly reduced the hydrogen economy to <10 kg H_2/100 km [87] and is one of the most significant results of the CHIC project in London. **Figure 12** shows that the fuel economy of the buses operated as part of the CHIC project showed considerable improvements over those in the CUTE project. It can also be seen that the London buses performed better than the CHIC target, exceeding it by nearly 50%. For all of the London FC buses, the hydrogen is stored as a compressed gas at 350 bar, with the gas cylinders stored on the roof of the bus [82].

Between 2011 and 2016, the FC buses in operation in London have covered over 1.1 million kilometres [89], and a number of the FC buses have achieved the milestone of 20,000 hours of operation [90]. This reflects the improvement of availability seen over the course of the deployment of CHIC's London fleet. **Figure 13** shows the availability from January 2012 until May 2015. The monthly availability of London FC buses has also significantly increased after

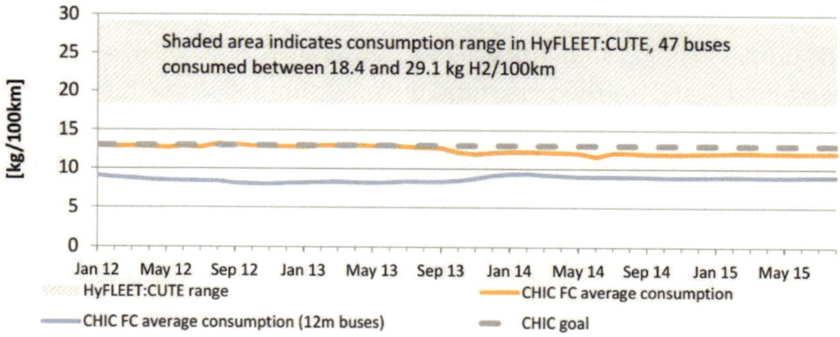

Figure 12. Average fuel consumption of FC buses in CHIC project (figure from FCH JU, 2016) [87].

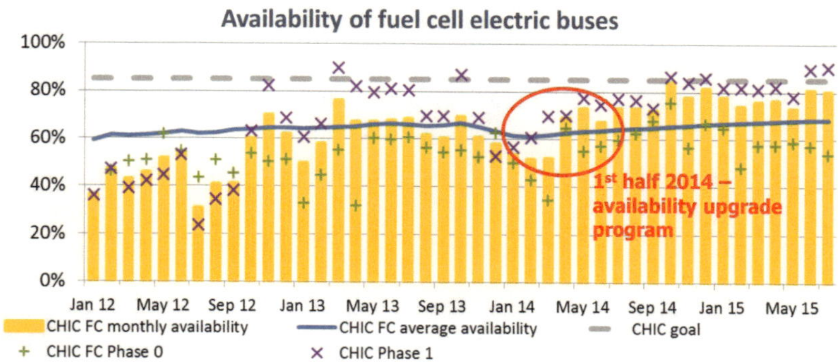

Figure 13. Availability of London FC buses in CHIC project (figure from FCH JU, 2016) [87].

the availability upgrade program carried out in 2014. The availability is expected to improve to over 85% by the end of the CHIC project as operators gain more operational and problem-solving experience.

Apart from the technical and economic improvements, the London trail buses have also proven that the technology became more viable because of the full working schedule, direct diesel replacement, centralised infrastructure and high public acceptance [86]. The trial test of FC-powered buses projects has provided promising performance as a long-term solution to zero emission transportation.

5. Comparison study

This part aims at to provide a comparison of the current state of low emission and zero emission bus systems. Diesel hybrid buses have been developed and deployed as a means of achieving emissions reductions, where a number of advantages in terms of efficiency, emissions and fuel consumption can be seen over diesel buses. There are, however, a number of problems associated with their widespread deployment. The first of these is the cost and is due to the additional components necessary for the electrical system. Second, the inclusion of the electrical system necessitates a significantly more complicated configuration [19]. Third, although diesel hybrid buses can offer significant improvements in terms of CO_2 and NO_x emissions, the primary energy source is still the ICE. As such, they fail to address the underlying source of emissions and are therefore fundamentally limited in the improvements that can be achieved. As such, they can only really be considered as a transitional technology to reduce emissions but are not a viable option for meeting zero emissions targets. In order to meet the requirements for zero emissions buses, which is the ultimate objective for a clean transportation network, technologies such as electric and FC buses have been developed as a long term solution for city bus transportation needs. Therefore, this section will mainly compare the battery electric bus (opportunity, overnight and trolley) and FC bus technologies as the two most promising zero emission solutions in terms of the operational requirements and is summarised in **Table 4**. The rankings are based on the authors' opinions with reasoning given in the paragraphs below.

Range: Opportunity e-buses have a smaller energy storage that requires frequent recharging, which equates to poor performance in terms of daily range. Overnight e-buses utilise a much larger battery, which increases the range with reported values of over 300 km per charge. Trolley e-buses are continuously powered with electricity by overhead lines along the route which effectively gives unlimited range. FC buses use hydrogen cylinders as the fuel tanks, which allow the range to be greatly extended (up to 450 km) for as much as hydrogen fuel cylinder weight and size allows [91].

Route flexibility: Opportunity and trolley e-buses require recharging infrastructures along the route which greatly limits their route flexibility. This is somewhat dependant on the size of the on-board battery and will likely be more acute for trolley e-buses. The overnight e-buses and FC buses are expected to be able to operate for an entire day's service without recharging

Zero emission option	Opportunity E-bus	Overnight E-bus	Trolley E-bus	Fuel cell bus
Daily range	4	3	1	2
Route flexibility	3	1	4	1
Refuelling time	2	3	Not available	1
Infrastructure	3	2	4	1
Fuel availability	1	1	1	4
Clean source	1	1	1	4
Cost	3	1	2	4

Notes: 1, best; 4, worst.

Table 4. High level comparison of operational performance of zero emission bus concepts.

or refuelling. As such this allows for much greater route flexibility. This appears to be easily achieved for FC buses, however for overnight e-buses this is not always the case and will again be dependent on the size of the battery.

Refuelling time: Opportunity e-buses require frequent recharging throughout the entire route. Although each recharges for the opportunity e-bus only takes up to 15 minutes, it is still considered as a drawback due to the requirement for regular recharging. Overnight e-buses require a longer recharging time (average >4 hours) after each operation due to the increased battery capacity. The recharging time is heavily dependent on the charging power. Trolley e-buses are charged through overhead wires so that they require no refuelling time. FC buses are refuelled with gaseous hydrogen, which can be completed quickly (<10 minutes) [91].

Infrastructure: Opportunity e-buses and trolley buses require corresponding infrastructure along the route and each end of the routes. Therefore, opportunity e-buses and trolley buses require a comprehensive infrastructure network. Overnight e-buses and FC buses both require infrastructure to recharge/refuel at the end of daily operation. This can, however, be centralised at the service depot and hence does not need to be as comprehensive. It appears, however, that the current recharging times for overnight e-buses presents a problem since it is likely that a significant number of recharging points and a massive recharging power would be needed to recharge the batteries of a large fleet in time for the next day's service. This could potentially be an issue for the electrical grid infrastructure if the number of buses grows significantly, while this would not be a problem for FC buses because of their short refuelling time.

Fuel availability: All three battery electric bus technologies use electricity to recharge their batteries. This electricity could be central managed and distributed locally through the local electricity grids; however, widespread electric bus deployment could significantly stress this infrastructure. FC buses will likely require the development of a comprehensive distribution network for hydrogen, although on-site hydrogen production has been demonstrated. Additionally, hydrogen fuel storage would also create additional cost.

Clean source: Real zero emissions bus technology needs to be clean throughout the manufacturing process, fuel production and bus operation. Currently, battery electric and FC bus technologies can achieve zero operating emission but the lifetime emissions are much harder to quantify. It is hard to forecast how the emissions from new technology manufacturing will change, but the fuel production method can be roughly estimated. In the UK, the GHG emissions for electrical energy were 0.44932 $kgCO_2/kWh$ in 2014 [92]. This is likely to change as the UK's energy mix changes, where in 2015, 24.6% of electricity was generated from renewable energy sources [93]. Similarly, for FC buses, the source of hydrogen is critical in determining the overall emissions. Currently, about 96% of hydrogen is derived from fossil fuels [94] which results in 13.7 $kgCO_2/kgH_2$ [95]. Despite this, investigations into the use of renewable energy for hydrogen production through the process of electrolysis have been carried out offering potential for a low carbon source of hydrogen. Currently, electricity for battery electric buses is a cleaner fuel than hydrogen for FC buses.

Cost: Both electric and FC buses have higher capital costs than a conventional diesel bus; however, FC buses are currently far more expensive than electric buses. The capital cost of electric buses is somewhat dependant on the type of operation expected, where overnight buses will have higher costs than opportunity and trolley buses due to the increased battery capacity. This does, however, need to be weighed up against the cost of infrastructure, where opportunity and trolley buses require a comprehensive and expensive charging network. Overnight electric and FC buses on the other hand can make use of a centralised recharging/refuelling infrastructure.

Throughout this chapter, the main technologies being implemented to meet the low emissions requirements have been presented. The most promising for these in terms of zero emissions are electric and FC buses; however, it is clear that there are still significant barriers to their widespread implementation. Following on from the challenges identified in the comparison section a number of challenges for future developments have been identified.

For electric buses, it is clear that further improvements to battery technology are required in terms of their energy densities and lifetime as well as the development of an effective charging infrastructure. The challenges are somewhat dependant on whether the bus is intended to use the overnight or opportunity charging schemes. For overnight charging, the charging infrastructure can be centralised; however, this necessitates very large power requirements for the charging infrastructure, additionally the range of the buses needs to be addressed through battery developments. The opportunity charging schemes a comprehensive and distributed charging network. In most cases, this requires the development of high efficiency and power wireless charging technologies.

The future development of FC buses requires development in a broader range of areas. This includes further work on individual components such as the FC stack and hydrogen storage. The FC stack is still the most expensive component of the FC bus. The further development of the control strategies for hybridised buses held significant promise in reducing the size of the required FC stack and improving the fuel economy. Hydrogen storage is a key area for future research for bus applications, where technologies such as solid state storage offer potential to improve the storage density of hydrogen. For widespread implementation, the development of the hydrogen infrastructure is vital. This includes the production of hydrogen, particularly from clean sources, the distribution of hydrogen or on-site production and purification.

Acknowledgements

We wish to thank the Engineering and Physical Sciences Research Council (EPSRC) who have funded the HyFCap (Reducing the cost and prolonging the durability of hydrogen fuel cell systems by in-situ hydrogen purification and technology hybridisation) (grant number: EP/K021192/1) jointly carried out by University College London and University of Sheffield.

Author details

Julius Partridge, Wei Wu* and Richard Bucknall

*Address all correspondence to: w.wu.11@ucl.ac.uk

University College London, London, UK

References

[1] Weston M, Emissions from the TfL bus fleet safety, accessibility and sustainability panel, Transport for London, 2015, pp. 11 http://content.tfl.gov.uk/sasp-20150707-part-1-item07-reducing-emmisisons-from-the-bus-fleet.pdf

[2] IEA. Energy and Air Pollution World Energy Outlook, Special Report. 2016, pp. 13-16

[3] Yamada H, Hayashi R,Tonokura K. Simultaneous measurements of on-road/in-vehicle nanoparticles and NOx while driving: Actual situations, passenger exposure and secondary formations. Science of Total Environment. 2015;**563-564**:944-955

[4] Degraeuwe B, Thunis P, Clappier A, Weiss M, Lefebvre W, Janssen S, Vranckx S. Impact of passenger car NO*x* emissions and NO_2 fractions on urban NO_2 pollution—Scenario analysis for the city of Antwerp, Belgium. Atmospheric Environnment. 2016;**26**(2):218-224

[5] Cooper E, Arioli M, Carrigan A, Jain U. Exhaust emissions of transit buses: Sustainable urban transportation fuels and vehicles. 2012;**40**(4):1-40

[6] TfL, Improving the health of Londoners, Transport action plan, greater London Authority, 2014, pp.32 http://content.tfl.gov.uk/improving-the-health-of-londoners-transport-action-plan.pdf

[7] TfL. Transport emissions roadmap. Transport London, 2014;**1**:1-48

[8] White P, Impacts of bus priorities and busways on energy efficiency and emissions, University of Westminster, 2015, pp.12 http://www.greenerjourneys.com/wp-content/uploads/2015/09/Binder2.pdf

[9] Robb A, The new bus for London - diesel/electric hybrid, Clean fleets case study, Clean fleets, 2014. pp. 1-5 http://www.clean-fleets.eu/fileadmin/New_Bus_for_London_Case_Study_for_Clean_Fleets_-_final.pdf

[10] Werkstetter, S, Ultracapacitor usage in wind turbine pitch control systems, Maxwell technologies white paper, 2015. pp.5 http://www.maxwell.com/images/documents/Wind_Turbine_Pitch_Control_White%20Paper_3000722_1.pdf

[11] Bush T, Eaton S, et al, Air pollution in the UK 2013, Department for environment food & rural affairs, 2014. https://uk-air.defra.gov.uk/assets/documents/annualreport/air_pollution_uk_2013_issue_1.pdf

[12] TfL, London bus fleet audit, Great London authorities, 2017, http://content.tfl.gov.uk/bus-fleet-audit-130117.pdf

[13] TfL, Ultra low emission zone update to the London Assembly, Cleaner Air for London, 2014, https://www.london.gov.uk/moderngov/documents/b9960/Minutes%20-%20Appendix%202%20-%20ULEZ%20Update%20Thursday%2006-Feb-2014%2010.00%20Environment%20Committee.pdf?T=9.

[14] Shawcross V. Bus services in London transport committee members. No. October, 2013.

[15] Coyle F, Manager E, Emissions T. London buses emissions reduction. Transport London, 2010;**1**:1-14.

[16] Plowden B, Ultra low emission zone (ULEZ) portfolio, Transport for London, 2015. http://content.tfl.gov.uk/board-20151217-pt1-item12-ulez.pdf

[17] German J. Hybrid-Powered Vehicles, 2nd Ed. London: SAE International; 2011.

[18] Herrmann F, Rothfuss F. Introduction to hybrid electric vehicles, battery electric vehicles, and off-road electric vehicles. In: Advances in Battery Technologies for Electric Vehicles, Elsevier, Woodhead Publishing Series in Energy; 2015. pp. 3-16.

[19] Folkson R, Conventional fuel hybrid electric vehicles. In: Alternative fuels and advanced vehicle technologies for improved environmental performance towards zero carbon transportation, Woodhead publishing series in energy, 2014, pp. 632-654. http://www.sae.org/images/books/toc_pdfs/BELS078.pdf

[20] Larminie J and Lowry J, Electric vehicle technology explained, 2nd ed, John Wiley & Sons, 2012. http://ev-bg.com/wordpress1/wp-content/uploads/2011/08/electric-vehicle-technology-explained-2003-j-larminie.pdf

[21] Brightman T, Girnary S, Bhardwa M, Bus idling and emissions, Passenger transport executive group, 2010, pp. 80. http://www.urbantransportgroup.org/system/files/PTEGBusIdling_ResultsReportfinalv10.pdf

[22] LCVP. The low emission bus guide. 2016. http://www.lowcvp.org.uk/initiatives/leb/Publications.htm

[23] Brown D, Hybrid buses - Surface transport panel, Transport for London, 2010, pp. 1-4. http://content.tfl.gov.uk/Item07-Hybrid-Buses-STP-30-june-2010.pdf

[24] Sevigny E, Urban transportation showcase program - annual reviews 2006-2008, Transport Canada, 2009. http://publications.gc.ca/collections/collection_2012/tc/T1-6-2008-eng.pdf

[25] Atkins P, Cornwell R, Tebbutt N, Schonau, Preparing a low CO_2 technology roadmap for buses, Ricardo LowCVP, 2013 http://www.apcuk.co.uk/wp-content/uploads/2015/10/LowCVP-Ricardo-Bus-Roadmap-FINAL.pdf

[26] Ling B, Gilbey, TfL hybrid bus monitoring, Low carbon vehicle partnership, 2010, pp. 11-14. http://www.lowcvp.org.uk/assets/workingdocuments/BWG-P-11-05%20TfL%20hybrid%20bus%20monitoring.pdf.

[27] TfL. New Routemaster buses on Route 453. 2014. [Online]. Available from: https://tfl.gov.uk/info-for/media/press-releases/2014/october/new-routemaster-buses-on-route-453

[28] Wright, Wright bus spec sheet, The new bus for London featuring hybrid electric driveline, Wright group, 2013, pp. 1-3. http://www.wrightbusinternational.com/datasheets/Routemaster%20spec%20sheet.pdf

[29] Mapelli F, Tarsitano D, et al. A study of urban electric bus with a fast charging energy storage system based on lithium battery and supercapacitor, Ecological vehicles and renewable energies, IEEE, 2013. http://ieeexplore.ieee.org/xpls/icp.jsp?arnumber=6521614

[30] Varga B, Iclodean C, Electric buses for urban transportation: assessments on cost, infrastructure and exploitation, Fascicle of management and technological engineering, 2015, pp. 253-258.http://imtuoradea.ro/auo.fmte/files-2015-v1/Bogdan%20Ovidiu%20VARGA%20-%20ELECTRIC%20BUSES%20FOR%20URBAN%20TRANSPORTATION%20-%20ASSESSMENTS%20ON%20COST,%20INFRASTRUCTURE%20AND%20EXPLOITATION.pdf

[31] Tie SF, Tan CW. A review of energy sources and energy management system in electric vehicles. Renewable and Sustainable Energy Review. 2013;**20**:82-102

[32] Yong JY, Ramachandaramurthy VK, Tan KM, Mithulananthan N. A review on the state-of-the-art technologies of electric vehicle, its impacts and prospects. Renewable and Sustainable Energy Review, 2015;**49**:365-385

[33] California Environmental Protection Agency Air Resources Board. Technology Assessment: Medium- and Heavy-Duty Battery Electric Trucks and Buses. No. October, 2015:**2015**;1-91

[34] Miles J, Potter S. Developing a viable electric bus service: The Milton Keynes demonstration project. Research in Transportation Economics. 2014;**48**:357-363.

[35] Bubna P, Advani SG, Prasad AK. Integration of batteries with ultracapacitors for a fuel cell hybrid transit bus. 2012;**199**:360-366

[36] Mahmoud M, Garnett R, Ferguson M, Kanaroglou P. Electric buses: A review of alternative powertrains. Renewable and Sustainable Energy Review. 2016;**62**:673-684

[37] Göhlich D, Kunith A, Ly T. Technology assessment of an electric urban bus system for berlin. WIT Transport Built Environment. 2014;**138**:137-149

[38] McKinsey. Urban buses: alternative powertrains for Europe, FCH JU, 2012. http://www.gppq.fct.pt/h2020/_docs/brochuras/fch-ju/20121029%20urban%20buses,%20alternative%20powertrains%20for%20europe%20-%20final%20report_0.pdf

[39] Mohamed A, Ayob A, Faizal WM, Mahmood W, Zamri M, Wanik C, Siam MM, Sulaiman S, Azit AH, Azrin M, Ali M. Review on electric vehicle, battery charger, charging station and standards. Research Journal of Applied Sciences Engineering and Technology. 2014;**7**(2):364-373

[40] Bloch-Rubin J, Gallo T, Tomic JB. Peak demand charges and electric transit buses. 2014;**5605**(626):744

[41] Singh AP. A new design of current collector for electric trolley bus. In: Transportation electrification conference, IEEE, 2015. http://ieeexplore.ieee.org/xpls/icp.jsp?arnumber=7386864

[42] IEA, Global EV outlook 2016 beyond one million electric cars, International energy agency, 2016. https://www.iea.org/publications/freepublications/publication/Global_EV_Outlook_2016.pdf

[43] Kailasam C. Strategic analysis of global hybrid and electric heavy-duty transit bus market, Frost & Sullivan, 2012. http://academy.busworld.org/lib/plugins/fckg/fckeditor/userfiles/file/seminars/busworld/in_the_footsteps_of_prime_minister_modi/12_chandramowli_kailasam_-_strategic_analyis_of_global_hybrid_and_electric_heavy_duty_transit_bus_market.pdf

[44] Zhou Y, Wang M, Hao H, Johnson L, Wang H, Hao H. Plug-in electric vehicle market penetration and incentives: a global review. Mitigation and Adaption Strategies for Global Change. 2015;**20**(5): 777-795

[45] Masiero G, Ogasavara MH, Jussani AC, Risso ML. Electric Vehicles in China: BTD Strategies and Government Subsidies. Statewide Agricultural Land Use Baseline. 2015;1:2015.

[46] Grutter J. Real world performance of hybrid and electric buses, Renewable energy & energy efficiency promotion in international cooperation, 2014. http://www.repic.ch/files/7114/4126/7442/Grutter_FinalReport_e_web.pdf

[47] Payne M. New generation transport: sub mode options investigation. Metro. 2014. www.ngtmetro.com/WorkArea/DownloadAsset.aspx?id=4294968292

[48] Brief TN. Chinese Electric Buses Enter Service in London. [Internet]. 2013. Available from: http://www.transportnewsbrief.co.uk/news/chinese-elelectric-buses-enter-service-london/

[49] BYD. 10.2m double deckor, RHD electric bus spec sheet. BYD group. 2015. http://www.bydeurope.com/downloads/eubs_specification/BYD_10_2_Meters_Electric_bus.pdf

[50] TfL. ZeEUS project London demonstration. Mayor of London. 2015. http://zeeus.eu/uploads/publications/documents/zeeus-london-leaflet.pdf

[51] Poulton M. Electric buses. No. July, 2011. www.triangle.eu.com/wp-content/.../LEVP-070714-Electric-Buses-Mark-Poulton.pdf

[52] Poulton M. Electric buses. London electric vehicle partnership meeting. Transport for London. 2014. http://unplugged-project.eu/wordpress/wp-content/uploads/2015/12/UNPLUGGED-Publishable-Final-Report.pdf

[53] ADL. BYD and ADL Partner to Supply Go-Ahead London with Capital's First, Large-scale Pure Electric Bus Fleet. [Internet]. 2015. Available from: http://www.alexander-dennis.com/news/byd-and-adl-partner-to-supply-go-ahead-london-with-capitals-first-large-scale-pure-electric-bus-fleet/

[54] Manufacturer T. First-ever Electric Double-decker London Red Bus. [Online]. Availablefrom: http://www.themanufacturer.com/articles/first-ever-electric-double-decker-london-red-bus/

[55] LowCVP. Market Monitoring: London's First Electric Double Decker. [Internet]. 2016. Available from: http://www.lowcvp.org.uk/initiatives/lceb/marketing/news,londons-first-electric-double-decker_3417.htm

[56] Andaloro L, Napoli G, Sergi F, Dispenza G, Antonucci V. Design of a hybrid electric fuel cell power train for an urban bus. International Journal of Hydrogen Energy, 2013;**38**(18):7725-7732

[57] Eshani M, Gao Y, Gay S, Emadi A. Modern Electric, Hybrid Electric and Fuel Cell Vehicles, CRC press. 2nd ed. 2010.

[58] Stoia T, Atreya S, O'Neil P. A Highly Efficient Solid Oxide Fuel Cell Power System for an All-Electric Commuter Airplane Flight Demonstrator. In: 54th Aerospace recherch central of AIAA association, 2016.

[59] Andersson M, Sunden B. Technology review - solid oxide fuel cell. Energiforsk. 2015. http://www.elforsk.se/Programomraden/El--Varme/Rapporter/?download=report&rid=2015_136_

[60] Steinberger WR. Study on the integration of an SOFC system into the on-board electricity system of the biogas bus. Baltic Biogas Bus. 2012. http://www.balticbiogasbus.eu/web/Upload/Use_of_biogas/Act_6_4/WP%206.4%20Final%20report_290812_ATI.pdf

[61] Barbir F, Introduction. In: PEM Fuel Cells, Academic press. 2nd ed., Elsevier; 2013. pp. 1-16.

[62] Villatico F, Zuccari F. Efficiency comparison between FC and ice in real urban driving cycles. International Journal of Hydrogen Energy. 2008;**33**(12):3235-3242

[63] Mekhilef S, Saidur R, Safari A. Comparative study of different fuel cell technologies. Renewable and Sustainable Energy Review. 2012;**16**(1):981-989

[64] Eberle U, von Helmolt R. Fuel cell electric vehicles, battery electric vehicles, and their impact on energy storage technologies: an overview. In: Electric and Hybrid Vehicles – Power Sources, Models, Sustainability, Infrasturcture and the Market. London: Elsevier; 2010. p. 242.

[65] Giorgi L, Leccese F. Fuel cells: Technologies and applications. Open Fuel Cells Journal. 2013;**6**:1-20

[66] Melo P, Ribau J, Silva, C. Urban bus fleet conversion to hybrid fuel cell optimal powertrains. Procedia - social and behavioral sciences. Volume 111. 2014. pp. 692-701. http://www.sciencedirect.com/science/article/pii/S1877042814001049

[67] Stempien JP, Chan SH. Comparative study of fuel cell, battery and hybrid buses for renewable energy constrained areas. Journal of Power Sources, 2017;**340**:347-355

[68] Element energy limited. Post-2014 London hydrogen activity: options assessment. Transport for London. 2014. http://www.hydrogenlondon.org/wp-content/uploads/2013/10/HydrogenBuses-Post-2014-221012.pdf

[69] Atmaja TD. PEMFC optimization strategy with auxiliary power source in fuel cell hybrid vehicle. The journal for technology and science. 2012;**23**(1):25

[70] Vural B, Boynuegri AR, Nakir I, Erdinc O, Balikci A, Uzunoglu M, Gorgun H, Dusmez S. Fuel cell and ultra-capacitor hybridization: A prototype test bench based analysis of different energy management strategies for vehicular applications. International Journal of Hydrogen Energy. 2010;**35**(20):11161-11171

[71] Bizon N. Energy efficiency of multiport power converters used in plug-in/V2G fuel cell vehicles. Applied Energy. 2012;**96**:431-443

[72] Sami BS, Abderrahmen BC, Adnane C. Design and dynamic modelling of a fuel cell/ultra capacitor hybrid power system. Electrical engineering and software applications. IEEE. 2013. http://ieeexplore.ieee.org/document/6578382/

[73] Pany P, Singh RK, Tripathi RK. Performance analysis of fuel cell and battery fed PMSM drive for electric vehicle application. Power, control and embedded systems. IEEE. 2012. http://ieeexplore.ieee.org/xpls/icp.jsp?arnumber=6508118

[74] Liukkonen M, Suomela J. Design of an energy management scheme for a series-hybrid powertrain. 2012 IEEE Transportion Electrification Conference and Expo, IEEE. 2012. pp. 1-6

[75] Hoffmann P. Tomorrow's Energy: Hydrogen, Fuel Cells, and the Prospects for a Cleaner Planet, The MIT press. 2002.

[76] Garcia P, Torreglosa JP, Fernandez LM, Jurado F. Control strategies for high-power electric vehicles powered by hydrogen fuel cell, battery and supercapacitor. Expert System Appliances. 2013;**40**(12):4791-4804

[77] Wu W, Bucknall RWG. Conceptual evaluation of a fuel-cell-hybrid powered bus. In: Proceedings of the University of Power Engineering Conference, IEEE. 2013. pp. 0-4

[78] Element energy. Strategies for joint procurement of fuel cell buses. The FCH JU. 2016. http://www.fch.europa.eu/sites/default/files/Strategies%20for%20joint%20procurement%20of%20FC%20buses_0.pdf

[79] IFCBC. All Active Demonstrations. [Internet]. 2016. Available from: http://gofuelcellbus.com/index.php/the-collaborative/all-active-demonstrations

[80] Eudy. L, Post M, Gikakis C. Fuel Cell buses in U.S. transit fleets: current status 2015. National renewable energy laboratory of U.S. department of energy. 2015. http://www.afdc.energy.gov/uploads/publication/fc_buses_2015_status.pdf

[81] National fuel cell bus program. Demonstrating advanced design hybrid fuel cell buses in Connecticut. National renewable energy laboratory of Federal transit administration. 2011. http://www.nrel.gov/hydrogen/pdfs/nfcbp_fs3_jul11.pdf

[82] HyFLEET:CUTE. Hydrogen transports - bus technology & fuel for today and for a sustainable future. European Union. 2009. http://gofuelcellbus.com/uploads/HyFLEETCUTE_Brochure_Web.pdf

[83] Air Products. Bringing hydrogen to London's streets. 2009

[84] 3Emotion. Van Hool Delivers Two Fuel Cell Buses for London. [Internet]. 2016. Available from: http://www.3emotion.eu/news/van-hool-delivers-two-fuel-cell-buses-london

[85] Ballard. Case study-fuel cell zero emission transit for the city of London. FCvelocity. 2016. http://ballard.com/files/PDF/Bus/TfL_Case_Study_Nov_2016.pdf

[86] Yorke D. The future is now an overview of the London hydrogen fuel cell bus project. First bus group. 2013. http://www.h2fcsupergen.com/wp-content/uploads/2013/06/The-Future-is-Now-London-H2-Bus-Project-David-Yorke-TfL.pdf

[87] CHIC. London hydrogen buses and the CHIC project. Fuel cells and hydrogen joint undertaking. 2016. http://www.all-energy.co.uk/RXUK/RXUK_All-Energy/2016/Presentations%202016/Hydrogen%20and%20Fuel%20Cells/Ben%20Madden.pdf?v=635993507410544891

[88] Tyler T, Core RD. Fuel Cell bus workshop. Bluways. 2011. http://gofuelcellbus.com/uploads/Bluways_IFCBW_2011.pdf

[89] Hydrogen London. London: a capital for hydrogen and fuel cell technologies. Mayor of London. 2016. https://www.london.gov.uk/sites/default/files/london_-_a_capital_for_hydrogen_and_fuel_cell_technologies.pdf

[90] Ballard. Fuel Cell electric buses; a solution for public transports. Ballard power systems group. 2016. www.h2fc-fair.com/hm16/images/forum/pdf/01monday/1240.pdf

[91] Berger R. Fuel Cell electric buses- potential for sustainable public transport in Europe. Fuel Cells and hydrogen joint undertaking. 2015. http://www.fch.europa.eu/sites/default/files/150909_FINAL_Bus_Study_Report_OUT_0.PDF

[92] BEIS. Government emission conversion factors for greenhouse gas company reporting. Department for business, energy & industrial strategy. 2016. https://www.gov.uk/government/collections/government-conversion-factors-for-company-reporting

[93] DUKES. Renewable sources of energy. In: Digest of United Kingdom energy statistics. Department for business, energy & industrial strategy. 2016. https://www.gov.uk/government/uploads/system/uploads/attachment_data/file/547977/Chapter_6_web.pdf

[94] WNA. Transport and the Hydrogen Economy. [Internet]. 2016. Available from: http://www.world-nuclear.org/information-library/non-power-nuclear-applications/transport/transport-and-the-hydrogen-economy.aspx

[95] Wu HJ. Parola VL. et al. Ni-based catalysts for low temperature methane steam reforming: recent results on Ni-Au and comparison with other Bi-Metallic systems. MDPI-Catalysts. 2013. http://www.mdpi.com/2073-4344/3/2/563/htm

Section 2

Technologies

Chapter 3

Advanced Charging System for Plug-in Hybrid Electric Vehicles and Battery Electric Vehicles

Muhammad Aziz

Additional information is available at the end of the chapter

http://dx.doi.org/10.5772/intechopen.68287

Abstract

The increase of plug-in hybrid electric vehicles (PHEVs) and battery electric vehicles (BEVs) results in higher electricity demand for their charging. In addition, the uncontrolled and timely concentrated charging is potential to decrease the quality of electricity. This condition has encouraged the development of advanced charging system, which is able to facilitate quick charging with minimum impacts on the electrical grid. This chapter explains some issues related to charging of PHEVs and BEVs including some available charging systems, charging behaviour and developed charging system employing battery for assistance during charging. In analysis of charging behaviour, the effect of ambient temperature to charging rate is clarified. Higher ambient temperature, such as during summer, leads to higher charging rate compared to one during winter. As advanced charging system, a battery-assisted charging system for PHEVs and BEVs is also described. The evaluation results in terms of their performance to facilitate a quick, simultaneous charging as well as reduce the stress of electrical grid due to massively uncontrolled charging are also provided. This system is considered as one of the appropriate solutions that can be adopted in the near future to avoid problems on electrical grid due to massive charging of PHEVs and BEVs.

Keywords: simultaneous charging, battery assistance, charging behaviour, charging rate

1. Introduction

Electric vehicles (EVs) have received an intensive attention during the last decade due to their characteristics as vehicles as well as other additional benefits that cannot be offered by conventional vehicles. A massive deployment of electric vehicles can reduce the total consumption of fossil fuel, therefore, cuts down the greenhouse gas emission [1]. In addition, as they have

INTECH
open science | open minds

© 2017 The Author(s). Licensee InTech. This chapter is distributed under the terms of the Creative Commons Attribution License (http://creativecommons.org/licenses/by/3.0), which permits unrestricted use, distribution, and reproduction in any medium, provided the original work is properly cited.

higher energy efficiency, lower running cost can be achieved than conventional internal combustion-engine vehicles. Recently, value-added utilization of electric vehicles also has been proposed and developed including the ancillary services for the electrical grid and electricity support to certain energy management system [2–5]. Therefore, the economic performance of the electric vehicles can be significantly improved.

Some literatures have proposed and described well the grid integration, especially the introduction of renewable energy, and electric vehicles [6]. The fluctuating renewable energy sources, such as wind and solar, require a fast-response energy buffer to cover their intermittency as well as and to store the surplus electricity due to higher supply side than demand side. Electric vehicles are considered as the appropriate resource to balance and store these kinds of renewable energy sources [7]. The battery owned by the electric vehicles can absorb and release the electricity from and to the electrical grid, respectively, to balance the electrical grid promptly.

In general, there are four types of electric vehicles currently running and developed: (i) conventional hybrid electric vehicle (HEV), (ii) plug-in hybrid electric vehicle (PHEV), (iii) battery electric vehicle (BEV) and (iv) fuel-cell electric vehicle (FCEV). HEV combines electric motor and internal combustion engine; hence, it is also fitted with a battery to power the motor as well as store the electricity. The energy to power the motor comes from the engine and regenerative breaking. However, recently, many HEVs have been redeveloped and shifted to PHEV due to the excellent characteristics and higher flexibility of PHEV than HEV. Like HEV, PHEV also owns electric motor and internal combustion engine.

According to IEEE standards, PHEV is HEV having following additional specifications: battery storage of larger than 4 kWh, charging system from external energy source and capability to run longer than 16 km [8]. Furthermore, BEV is generally defined as the vehicle driven solely by electric motors and the source of electricity is stored and converted from chemical energy in the battery. Therefore, BEV relies on external charging and its driving range depends strongly on its battery capacity. As the battery capacity of BEV is significantly larger than HEV and PHEV, battery makes up a substantial cost of BEV. Advanced development of battery and decrease of its price is highly expected in the near future; hence, more massive deployment of PHEVs and BEVs can be realized.

On the other hand, FCEV uses only electric motor like BEV. However, it utilizes hydrogen as the main fuel that is stored in the tank. The oxidation of hydrogen produces electricity to power the electric motor and if there is any surplus it is stored in the battery. In practice, as the hydrogen refuelling can be performed in a very short time, almost similar to one of the gasoline refuelling, FCEV basically facilitates no charging from the external charger.

Although it varies, the battery capacity of PHEV is generally larger than HEV. According to survey conducted by Union of Concerned Scientists (UCS), about 50% of drivers in US drive less than 60 km on weekdays [9]. Therefore, many available PHEVs can hold for a weekday commuting without additional charging outside. In addition, although its battery capacity is lower than BEV, PHEV has higher flexibility on driving range as the power can be supplied by the engine once the battery capacity drops to certain low value. Both PHEV and BEV are believed will dominate the share of vehicles in the future. In addition, according to Electric Power Research Institute (EPRI), around 62% of vehicles will encompass of PHEVs [10].

High share of PHEV and BEV results in high demand of electricity due to charging; hence, it strongly correlates to the supply and balancing of electrical grid.

Unmanaged charging of PHEVs and BEVs potentially results in several grid problems including over and under voltage and frequency in distribution networks, especially when individual charging of PHEVs and BEVs takes place in large number and capacity [11]. Some methods to minimize the impact of unmanaged charging of PHEVs and BEVs have been proposed and developed by some researcher. They include coordinated charging [12], demand response [13], battery-assisted charging [14] and appropriate charger distribution [15]. In addition, an integrated vehicle to grid (V2G) is also potential to avoid the concentrated charging, as well as facilitate the other services [16].

In the coordinated charging, the charging behaviour of PHEVs and BEVs are controlled by certain entities; therefore, the electrical grid can be maintained stable and balance. Further, this charging behaviour control is then correlated strongly with the V2G services, especially for load-shifting or valley filling strategy [17]. However, the algorithm for valley filling under large-scale vehicles deployment is very sophisticated; hence, computational complexity becomes a very crucial factor [18].

Demand response encourages the users or drivers of PHEVs and BEVs to manage their charging demand during peak-load hours or when the electrical grid system is at risk [19]. Therefore, it is usually divided into two types: time-based and incentive-based. The former deals strongly with the real-time pricing and critical peak pricing. On the other hand, the latter is related to the incentive due to utilization of PHEVs and BEVs for frequency regulation and spinning reserve [20]. Pricing system in the electrical grid requires accurate prediction on both supply and demand sides. Therefore, the uncertainties clarification and their impacts minimization become the major concern in demand response.

Although they are promising methods, both coordinated charging and demand response require further theoretical developments and demonstrations on to ensure the system and standard in a relatively massive control system. On the other hand, the battery-assisted charging is considered simple and applicable, due to its simplicities and convenience in structure and control.

This chapter discusses the charging system for both PHEV and BEV including the recently developed battery-assisted charger. At the beginning, available charging levels and systems for PHEVs and BEVs are explained initially in terms of charging rate and standards. In addition, the charging behaviour of the PHEV and BEV in different ambient temperature (seasons) are also described, clarifying the effect of ambient temperature to the charging rate. At the last, an advanced charging system with battery assistance is also explained including their quick-charging performance during simultaneous charging of electric vehicles.

2. Charging system for PHEV and BEV

Charging of PHEVs and BEVs correlates strongly with some parameters including charging devices, cost, charging rate, location, time and grid condition. Therefore, relevant selection and distribution of chargers are very crucial to be able to accommodate those parameters appropriately. PHEV and BEV basically share the same charging standards; therefore, there is no peculiar charger features or requirements for each vehicle. Charger is designed to be able to communicate with the vehicle to ensure the safety and appropriate electricity flow. In addition, charger also monitors the earth leakage at the surrounding ground.

On the other hand, battery management system (BMS) is installed in the vehicle as a very vital component, which is performing a thermal management, cell balancing and monitoring of over-charge and discharge of the battery pack. The battery pack consists of many individual cells having certain safe low working voltage. Therefore, it is very crucial to ensure that they are operating within the permitted range to avoid shorter battery life and battery failures, including fire.

Chargers can be installed on-board and off-board. The on-board charger limits its electricity flow because of some constraints, such as weight and space. It can be performed though conductive (direct contact through charging connector and cable) and inductive ways (using the electromagnetic field). On the other hand, the off-board charger is installed externally; therefore, there is no limitation related to size and weight. The electricity flow from the charger to vehicle is a DC flow; hence, high charging rate can be achieved.

The direction of electricity between charger and vehicle can be classified into unidirectional and bidirectional flows. The former only facilitates a single direction charging from external charger to the vehicle (battery). The latter provides the possibility of charging and discharging the electricity to and from the vehicle. Through bidirectional charging the utilization of PHEVs and BEVs is greatly widen.

Correlated to the charging rate, chargers or electric vehicle supply equipment (EVSEs) can be classified by its maximum amount of electricity possibly charged to the battery of PHEV or BEV, as follows:

a. Level-1 charging

Level-1 charging utilizes the on-board charger and is compatible with the household electrical socket and power, which generally has voltage of 100 or 200 V (AC) depending on the region. This level of charging can facilitate charging rate up to about 4 kW. This level of charging is suitable for the overnight charging at the ordinary household without the need of additional device installation.

b. Level-2 charging

This level of charging has the purpose of improving the charging rate by using the dedicated mounted-box. This level-2 charging can supply power of 4–20 kW, with a maximum voltage of 400 V (AC three phase), depending on the available capacity of local supply. Generally, this kind of chargers is installed at dedicated charging facilities including residential areas or public spaces. The charging connectors for both level-1 and level-2 chargers vary across the countries and manufactures.

c. Level-3 charging

Different to the above levels of charging, level-3 charging is performed in DC system. DC electricity is supplied by the charger, bypassing the on-board charger. Therefore, very high charging rate, higher than 50 kW, can be achieved. Currently, there is no single standard for this kind of fast charging which is accepted by all vehicle manufactures. The charging plug (including EV socket) and the communication protocol between the charger and vehicles are different between the standards, although the basic principles are similar.

Currently, there are three major standards of charger, especially for quick charging: CHAdeMO, combined charging system (CCS) and Tesla Supercharger. The detailed specifications of each charging standard are shown in **Table 1**.

CHAdeMO was the first, DC fast charging standard originally developed by Japanese companies including Tokyo Electric Power Companies (TEPCO), Fuji Heavy Industries, Nissan, Mitsubishi Motors and Toyota, which are organized by CHAdeMO Association. CHAdeMO standard also complies with international standard of IEC 62196-3. This standard is designed only for DC fast charging. According to the development roadmap [21], high power CHAdeMO is also developed with which is able to charge with 100 kW continuous power and 150–200 kW peak power (350 A, 500 V). In addition, further higher power CHAdeMO is also planned in future (2020) which can charge with charging rate of 350–400 kW (350–400 A, 1 kV). Currently, CHAdeMO has the largest global coverage, including Japan (about 7000 chargers), Europe (about 4000 chargers) and USA (about 2000 chargers).

Properties	**CHAdeMO**	**Combo 1 (CCS)**	**Combo 2 (CCS)**	**Tesla supercharger**
DC charging				
Max. voltage (V)	500	600	850	480
Max. current (A)	125	150	200	
Connector	CHAdeMO	Combo 1 (IEC 62196-3/SAE J1772)	Combo 2 (IEC 62196-3)	Special
Max. power (kW)	50	90	170	120
AC charging				
Voltage (V)		250	400 (3 phase) 230 (1 phase)	
Current (A)		32	63 (3 phase) 70 (1 phase)	
Max. power (kW)		13	44	
Others				
Charging signal	CAN	PLC		
Charging protocol	CHAdeMO	HomePlug Green Phy		Special

Table 1. Specification of charging standards for DC fast charging.

On the other hand, CCS standards, including Combo 1 and 2, are capable to facilitate both AC charging, including level-1 and level-2 charging, and DC charging. It was developed by several European and US car manufactures in around 2012. Society of Automotive Engineers (SAE) and European Automobile Manufacturer's Association (ACEA) strongly supported this initiative with the main purpose of facilitating both AC and DC charging with only single charging inlet in the vehicle. CCS is able to facilitate AC charging at maximum charging rate of 43 kW and DC charging at maximum charging rate of 200 kW with the future perspective of up to 350 kW [22]. CCS chargers are currently installed mainly in Europe and the USA with approximate numbers of 2500 and 1000, respectively.

Tesla Supercharger uses its own charging standard. Currently, Tesla Supercharger includes multiple chargers that are working in parallel and able to deliver up to 120 kW of DC charging [23]. Tesla Superchargers are currently installed in about 800 stations, having about 5000 superchargers in total.

Other charging method for PHEV and BEV includes inductive charging, which is conducted wirelessly. The electromagnetic induction is created by the induction coil, which is charged with high-frequency AC. The generated magnetic field will induce the vehicle-side inductive power receiver; thus, the electricity can be transferred to the vehicle. Inductive charging uses the family of IEC/TS 61980 standards. The application of inductive charging is potential to eliminate the range anxiety, as well as reduce the size of battery pack. However, there are some technical barriers in its application, especially related to lower efficiency, slower charging rate, interoperability and safety.

3. General charging behaviour of electric vehicles

In general, PHEVs and BEVs adopt lithium-ion battery for energy storage due to high energy density, longer charging and discharging cycles, lower environmental impacts and more stable electrochemical properties [24]. In general, charging and discharging of lithium-ion batteries are greatly influenced by the temperature. According to literatures [25, 26], lower rates of charging and discharging occur under relatively lower temperature. This is due to the change of interface properties of electrolyte and electrode such as viscosity, density, dielectric strength and ion diffusion [27]. Furthermore, the transfer resistance also increases, which could be higher than the bulk and solid-state interface resistances, as the temperature decreases [28].

Aziz et al. [14] have performed a study to clarify the influence of ambient temperature or season to charging rate of PHEV and BEV. The study was performed during both winter and summer, using CHAdeMO DC quick charger having rated power output of 50 kW. In addition, Nissan Leaf having battery capacity of 24 kWh was used as the vehicle. The results of their study are explained below.

Figure 1 shows the obtained charging rate and battery state of charge (SOC) under different seasons. Although the rated output capacity of the quick charger is 50 kW, the realized charging rate to vehicle is lower, especially during winter. Charging during summer (higher ambient temperature) leads to higher charging rate; therefore, shorter charging time can be

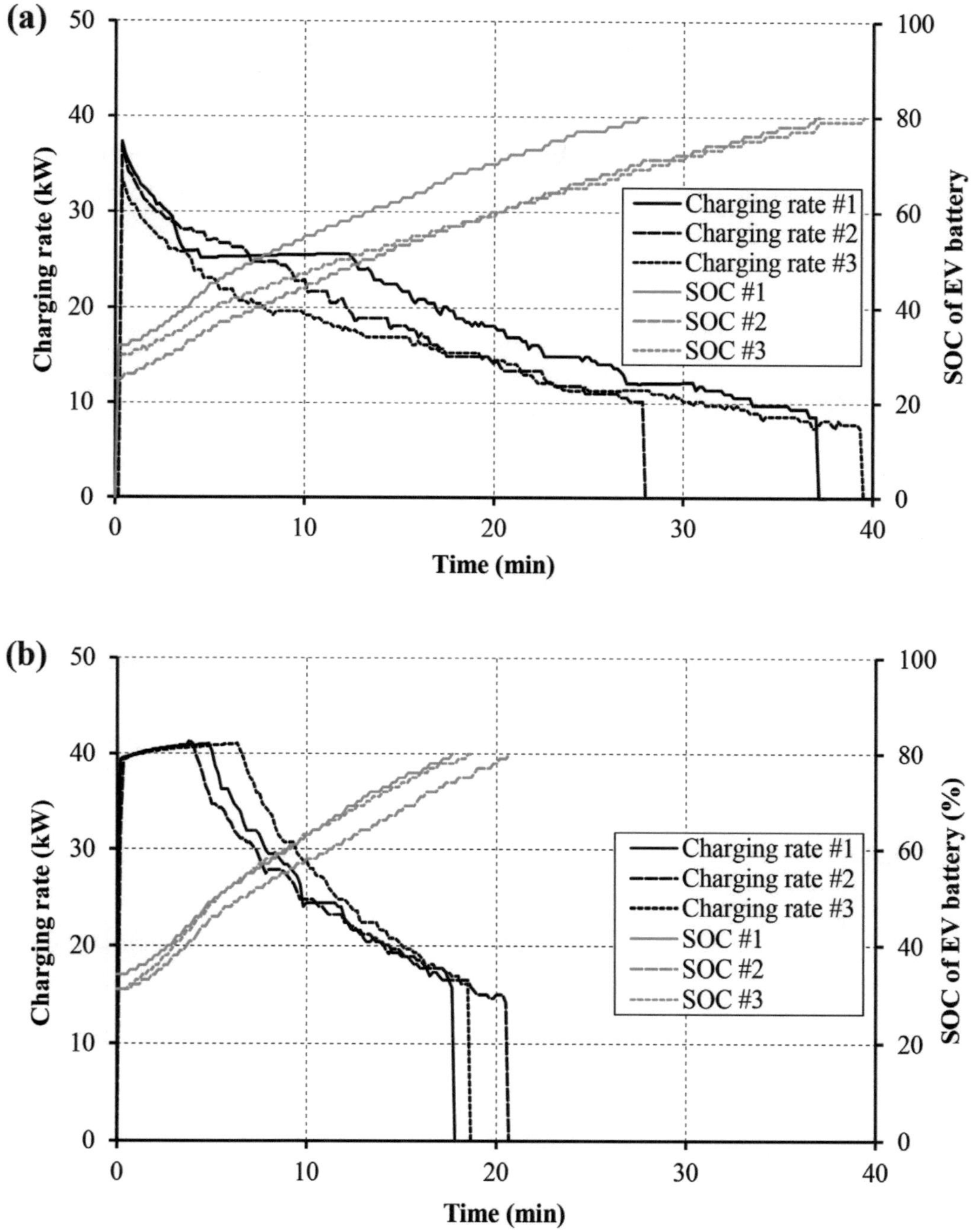

Figure 1. Charging rate and SOC change of battery in different season of charging: (a) winter, (b) summer.

achieved. To charge to battery SOC of 80% from about SOC of 30%, the required charging durations in both winter and summer are 35 and 20 min, respectively. During summer, a relatively high charging rate (about 40 kW) can be achieved up to an SOC of about 50%. However, the charging rate decreases moderately in accordance with the increase of battery

SOC. The charging rate at battery SOC of 80% is about 16 kW. On the other hand, during winter, the charging rate reaches about 35 kW instantaneously in relatively short duration and then decreases following the increase of battery SOC. In addition, the charging rate at battery SOC of 80% is about 10 kW.

Figure 2 shows both current and voltage changes during charging under different seasons. The curves of charging current are almost similar to charging rates in **Figure 1**. Lithium-ion

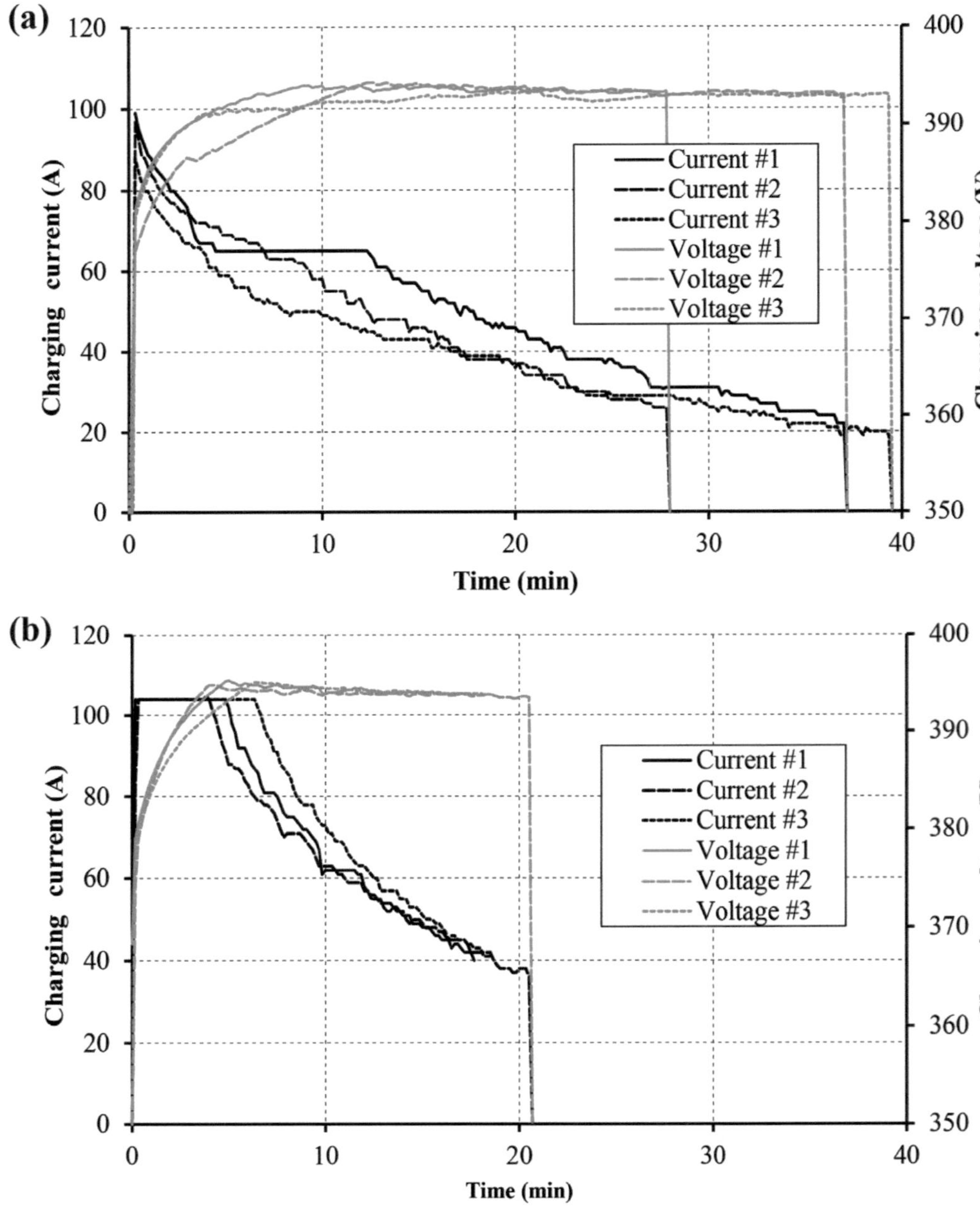

Figure 2. Current and voltage during charging in different seasons: (a) winter, (b) summer.

http://dx.doi.org/10.5772/intechopen.68287

batteries are generally charged with a constant current (CC)–constant voltage (CV) method [22]. Charging under lower temperature leads to a gradual decrease in the charging current with charging time or increase in battery SOC. In contrast, charging under relatively warmer conditions resulted in a higher charging current, especially at low battery SOCs. Higher CC of about 105 A is obtained at the initial charging of 5–10 min (battery SOC of up to about 50%). With regard to charging voltage, although there is no significant difference between charging in both conditions, charging in a relatively higher temperature (summer) results in a higher initial charging voltage before it is settling down to a certain constant value. Therefore, the CV condition can be reached faster.

It is clear that the ambient temperature affects significantly the charging behaviour of PHEVs and BEVs. Charging under relatively high ambient temperature (such as summer) facilitates a higher charging rate, especially because of higher charging current and faster increase in the charging voltage. Hence, a shorter charging time can be achieved.

When the vehicles are near to empty, the electricity can flow at a high rate and it starts to pace down when the battery SOC is higher than 50%. In addition, it gets really slower when SOC is higher than 80%. This phenomenon is generally called as tapering.

4. Advanced charging system

The widespread deployment of PHEV and BEV charging, especially fast charging, has some critical impacts on the electrical grid including the quality deterioration of the grid and grid overload. Therefore, it is very crucial to schedule and control the charging of PHEVs and BEVs. One strategic method to charge the vehicles with minimum impact on the electric grid is to adopt a battery to assist the charging. Aziz et al. [14] have proposed and studied the battery-assisted charger (BAC) for PHEV and BEV. The battery is embedded inside the charger with the aims of improving the quick-charging performance and minimizing the concentrated load to the grid.

The developed BAC is able to limit the received power from electrical grid, as well as control the charging rate to the vehicles. It is important to manage the received power from the grid in order to avoid the electricity demand larger than the contracted capacity and also optimize the electricity demand following the grid conditions. In future, as the share of renewable energy increases, the electrical grid also faces some problems including intermittency. This leads to the requirements of energy storage and demand control.

BAC manages the electricity distribution inside the system, such as electricity received from the grid, battery and chargers, to realize the optimum performance. Therefore, BAC is able to satisfy both supply side (minimizing the grid load through load shifting and reduction of electricity cost) and demand side (fascinating the vehicle owners through quick charging, although during peak hours).

The purposes of BAC covers: (1) reducing the contracted power capacity from the electrical grid, (2) avoiding the high electricity demand during peak hours due to PHEV and BEV charging, (3) shortening the charging time, as well as the waiting/queueing time, (4) facilitating a possible participation to the grid-ancillary programs such as spinning reserve and frequency regulations, (5) facilitating as storage for surplus electricity in the electrical grid due to high generated power by renewable energy and excess power and (6) providing an emergency back-up to the surrounding community in which it is installed.

Figure 3 represents the schematic diagram of the proposed BAC (solid and dashed lines serve both electricity and information streams, respectively). A community energy management system (CEMS) correlates to the whole management of energy throughout the community, covering supply, demand and storage. It monitors and controls the energy inside the community to ensure the comfort and security of community members as well as minimize the environmental influences and social cost. Concretely, CEMS communicates with other EMSs under its authority including electricity price and supply and demand forecast. In addition, it also negotiates with other CEMS or utilities outside the community to achieve the largest benefits for the whole community.

In the electricity stream, there are three main components that are connected by high-capacity DC lines: 1) AC/DC inverter, 2) stationary battery for storage and buffer and 3) quick charger for vehicles. The AC/DC inverter receives the electricity from electrical grid and converts it to

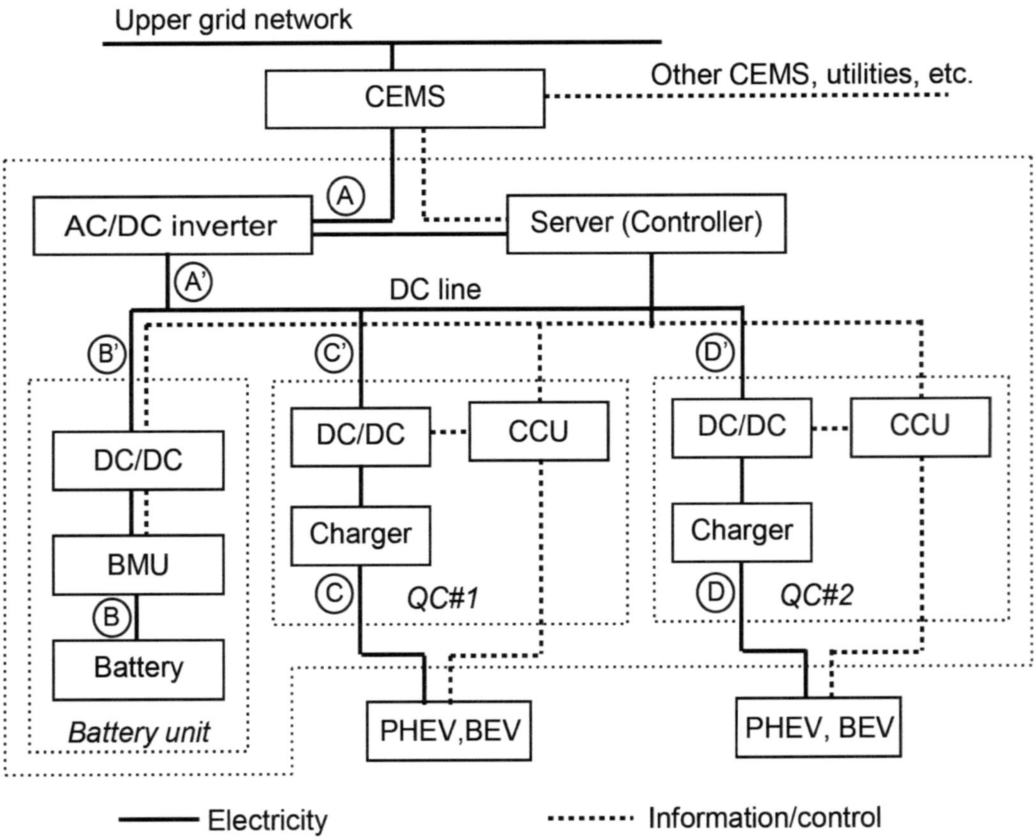

Figure 3. Schematic diagram of BAC system for PHEV and BEV.

relatively high voltage DC, which is about 400 V. In addition, the server/controller monitors, calculates and controls the amount of electricity received from the electrical grid based on some data, including electricity price and grid condition. Furthermore, the server manages the electricity to and from the battery and the charging rate from a quick charger to the connected vehicles. In the battery unit, a bidirectional DC/DC converter and battery management unit (BMU) are introduced to facilitate controllable charging and discharging behaviours according to the control values from the server. In the quick charger, a DC/DC converter and a charging control unit (CCU) are introduced to facilitate active management during vehicle charging. The number of quick chargers can be more than one, depending on the conditions.

The battery is adopted to store the electricity in case of the presence of remaining contracted power capacity and lower electricity price (during off-peak hours). In addition, the battery discharges its stored electricity in case of high electricity price due to high demand for charging or peak hours. The stationary battery having relatively large capacity is generally employed to sufficiently facilitate simultaneous charging of multiple vehicles. Therefore, the charging service can be maintained with high quality.

According to the charging and discharging behaviours of the employed stationary battery and the source of electricity for charging, quick-charging modes of the BAC are classified as follows:

a. Battery discharging mode

Stationary battery releases its electricity to assist the charging. Therefore, vehicle charging is conducted using electricity received from the electrical grid and discharged from the stationary battery. This mode is introduced when a simultaneous quick charging of multiple vehicles occurs, especially in case of high electricity price. Electricity in the battery discharging mode can be shown as follows:

$$P_{grid} + P_{bat} = P_{QC1} + P_{QC2} + P_{loss} \tag{1}$$

where P_{grid}, P_{batt}, P_{qc} and P_{loss} are electricity received from electrical grid, charged (negative value) or discharged (positive value) electricity from stationary battery, discharged electricity for quick charging of vehicles and electricity loss, respectively.

b. Battery charging mode

When there is remaining electricity (margin between the contracted power capacity and the used electricity) or the electricity prices is getting down (because of surplus electricity in the grid, night time, etc.), the stationary battery is charged to store electricity. The flow in this mode is expressed as Eq. (2).

$$P_{grid} - P_{bat} = P_{QC1} + P_{QC2} + P_{loss}. \tag{2}$$

c. Battery idling mode

Stationary battery might be in the idling (stand-by) mode in case of several conditions: (a) contracted power capacity can sufficiently cover the electricity demand for simultaneous charging of vehicles (low charging demand), (b) stationary battery is empty or under certain threshold value due to high and continuous charging of vehicles (stationary battery is not able to supply the electricity unless being recharged). In the latter, BAC manages the charging rate of each charger to corresponding vehicle; hence, the contracted power capacity can be maintained avoiding any penalty. Electricity flow in the battery idling mode can be represented as follows:

$$P_{grid} = P_{QC1} + P_{QC2} + P_{loss}. \tag{3}$$

BAC always maintains that the value of P_{grid} must be lower than or maximally equal to the contracted power capacity. Furthermore, P_{loss} is the total power loss in the system due to some factors, such as AC/DC and DC/DC conversions and internally consumed electricity the system. Therefore, the value of P_{loss} in each quick-charging mode might be different.

Table 2 shows the specification of the developed BAC system and the used vehicles during experiments. Nissan Leaf having battery capacity of 24 kWh is used as the vehicle. **Figure 4** shows the results of simultaneous quick charging of two vehicles during winter conducted using conventional quick charger and BAC under contracted power capacity of 50 kW. The electricity received from the grid is kept at 50 kW or below. In case of charging using the conventional charging system, the first connected vehicle is charged with higher charging rate than the vehicles connected later. This is due to the limit on contracted power capacity as well as the available power for charging. The charging rate of the second connected vehicle increases gradually as the charging rate of the first connected vehicle starts to decrease; therefore, the total electricity can be maintained to be lower or equal to the contracted power

Component	Property	Value
Installed battery in the charger	Battery type	Li-ion
	Total capacity	64.2 kWh
	Nominal voltage	364.8 V
	Maximum charging voltage	393.6 V
	Cut-off voltage during discharge	336.0 V
	Maximum current during discharge	176 A
	SOC threshold during charge	90%
	SOC threshold during discharge	10%
Quick charger	Number	2 units
	Standard	CHAdeMO
	Output voltage	DC 50–500 V
	Output current	0–125 A
	Rated output power	50 kW
Vehicle	Vehicle type	Nissan Leaf
	Battery type	Laminated li-ion
	Total battery capacity	24 kWh
	Maximum voltage	403.2 V
	Nominal voltage	360 V
	Cell rated capacity	33.1 Ah (0.3 C)
	Cell average voltage	3.8 V
	Cell maximum voltage	4.2 V

Table 2. Specifications of the developed BACS.

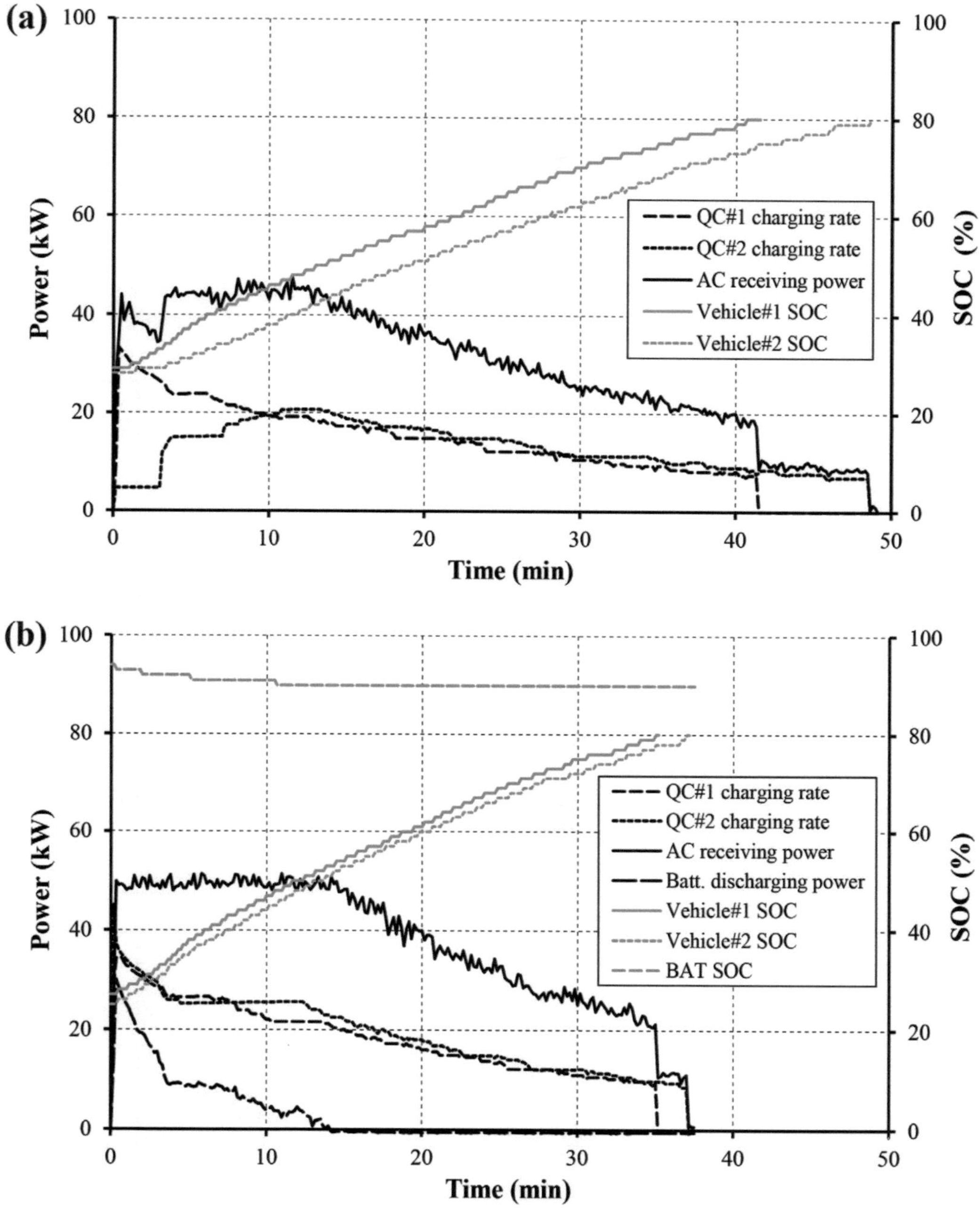

Figure 4. Charging of two BEVs during winter with different charger: (a) conventional charger, (b) developed BAC.

capacity. In addition, when the charging rate of both the connected vehicles decreases due to an increase of battery SOC, the total electricity purchased from the electrical grid decreases. The first and second vehicles are charged to SOC of 80% after charging for 40 and 50 min, respectively.

In contrast, in case of charging using the BAC, the first and second vehicles can enjoy almost the same charging rate, and both vehicles reach battery SOC of 80% in almost the same time (about 35 min). Furthermore, the electricity from electrical grid can be kept below the contracted power capacity, although the total charging rate for both vehicles is larger than the contracted power capacity. This is because the battery assisting the system was discharged to supply electricity. Hence, compared to a conventional charging system, BAC is able to achieve high-quality charging with higher charging rate during simultaneous charging.

Figure 5 shows the results of simultaneous charging of two vehicles during summer performed using conventional charging system and BAC. A same tendency with charging during

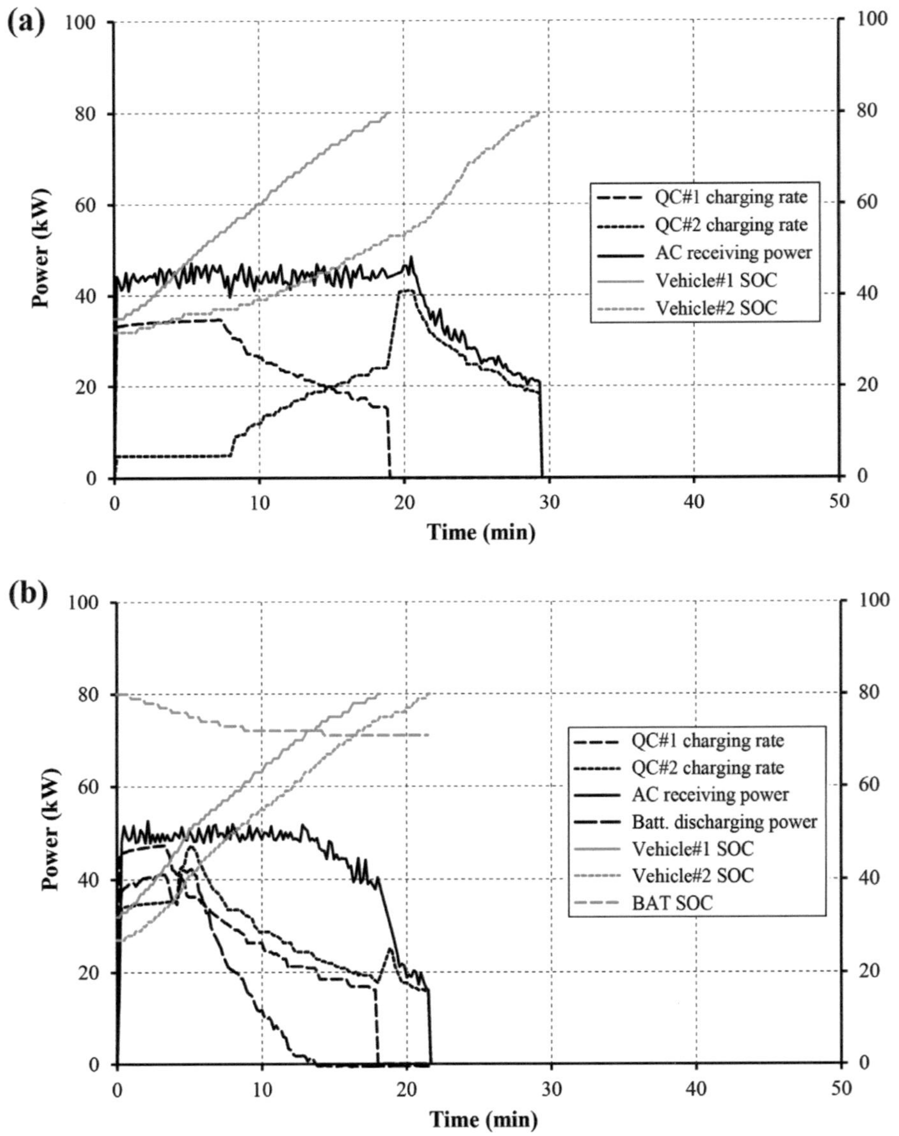

Figure 5. Charging of two BEVs during summer with different charger: (a) conventional charger, (b) developed BAC.

http://dx.doi.org/10.5772/intechopen.68287

winter, in the conventional charging system, the first connected vehicle enjoys a higher charging rate, while the second vehicle must contend with significantly lower charging rate because of limited contracted power capacity. The first and second vehicles reach battery SOC of 80% after charging of about 20 and 30 min, respectively.

When charging with BAC, similar to the case in winter, both vehicles could be charged almost at the same charging rate while maintaining the contracted power capacity. Both vehicles could be charged in a relatively short time of about 20 min. The stationary battery discharges its electricity until the total charging rate of two vehicles is equal or lower than the contracted power capacity.

It is clear that BAC improves the charging quality, especially during simultaneous charging of multiple vehicles. In addition, from the point of view of the electricity grid, application and deployment of BAC can reduce the stress on the grid because of the high demand for vehicle charging.

5. Simultaneous charging with developed BAC system

Figure 6 shows the demonstration test results during winter and summer under the contracted power capacity of 30 kW. Simultaneous charging of eight vehicles during summer can be conducted quicker than one during winter because of higher charging rate. However, the SOC of the stationary battery decreases considerably. It is because of the high discharging rate of the stationary battery to assist the quick chargers as well as cover the electricity demand due to limit of the contracted power capacity. In addition, the stationary battery cannot be charged because of no available marginal electricity from the electrical grid.

On the other hand, the discharging rate of the stationary battery is significantly lower during winter due to slower charging rate to the vehicles. Hence, the total charging rate of two quick chargers can be maintained to be lower than the contracted power capacity. It results in the marginal electricity that can be utilized to charge the stationary battery. Therefore, the SOC of the stationary battery in winter does not largely decrease compared to one during summer.

Figure 7 shows the simultaneous charging of eight vehicles during summer under a contracted power capacity of 15 kW. Compared to **Figure 6**, there is almost no significant change in the charging rate of vehicles, except that of the last connected vehicle. However, the discharging rate of the stationary battery is very high, resulting in significant decrease in its SOC. The SOC of stationary battery drops rapidly and reaches 10% during charging of the last two vehicles. As the result, the last connected vehicle is charged only using the electricity received from the electrical grid, with no assistance from the stationary battery. As the contracted power capacity is very low, the very last connected vehicle is not charged until the vehicle before it is charged completely. The stationary battery cannot be charged during simultaneous charging because of the lack of marginal electricity and the high charging rate of vehicles.

Based on the results of the demonstration test, the application of BAC is potential to improve significantly the charging performance of quick chargers, especially during the simultaneous charging of multiple vehicles. The balance among vehicle charging rate, contracted power capacity and stationary battery SOC seems to be very important. Therefore, PHEVs and BEVs charging demand must be forecast initially.

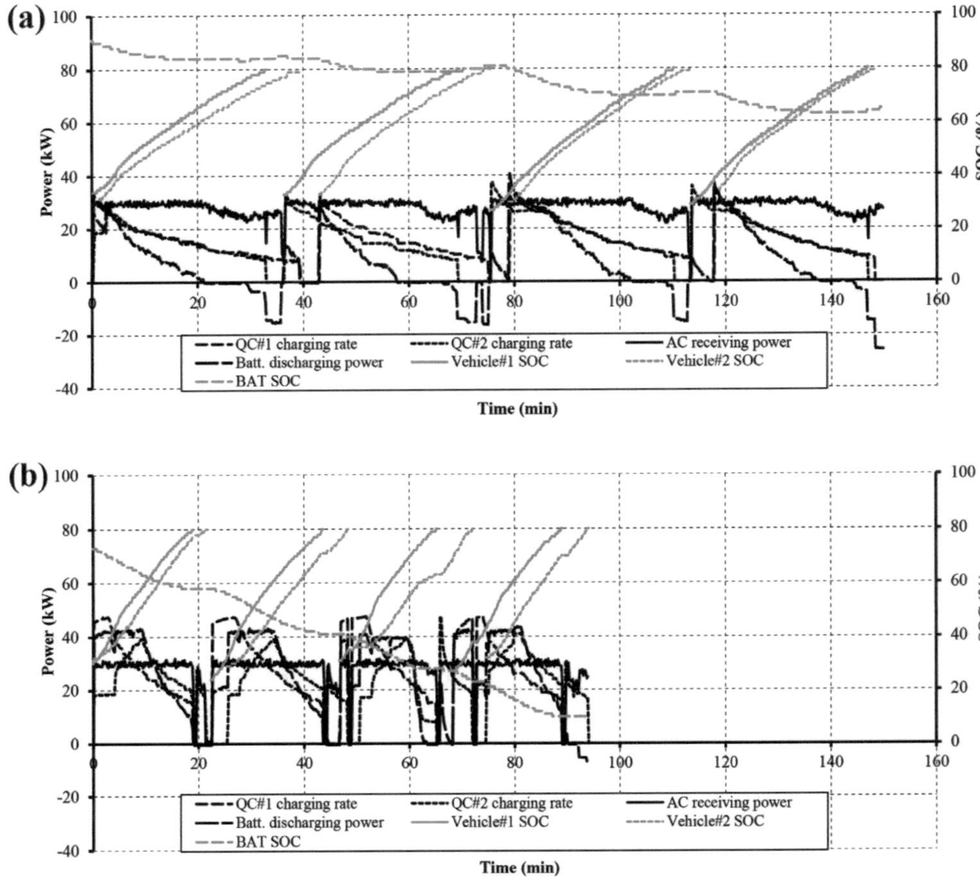

Figure 6. Simultaneous charging of eight vehicles using BAC under contracted capacity of 30 kW: (a) winter, (b) summer.

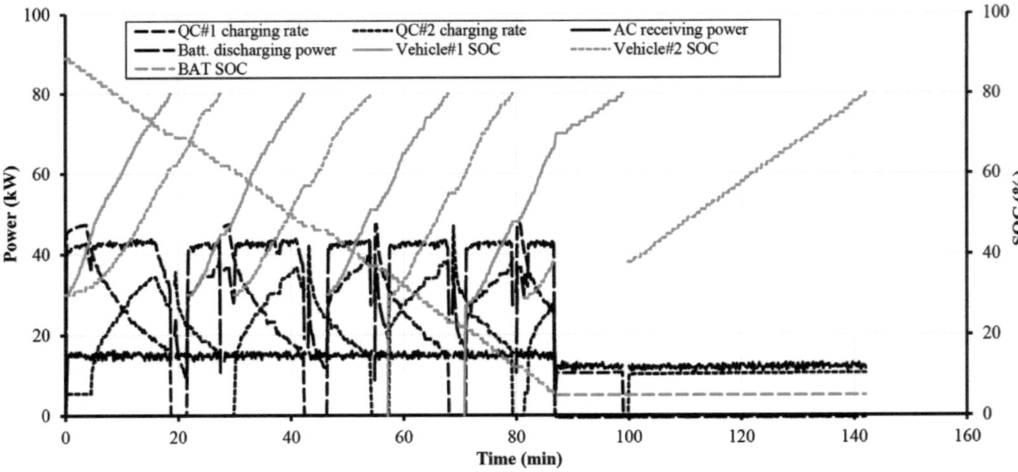

Figure 7. Simultaneous charging experiments of eight vehicles using developed BAC under contracted capacity of 15 kW during summer.

6. Conclusion

As the number of PHEVs and BEVs is massively increasing, their charging becomes a very important issue due to fluctuating and high demand of electricity. Therefore, it is very important to manage their charging through coordinated charging, battery-assisted charging and demand respond. Among these three methods, coordinated charging and demand respond require advanced theoretical development, massive demonstration and coordination in the electrical grid, therefore, they need couple of years in the future for realization. On the other hand, battery-assisted charging is considered very applicable in terms of economy and technology.

Charging behaviour of PHEV and BEV is strongly influenced by ambient temperature. Charging under relatively high ambient temperature (summer) leads to higher charging rate; therefore, shorter charging time can be realized. In addition, battery-assisted charger (BAC) has been developed especially to facilitate simultaneous charging of multiple vehicles under certain limited contracted power capacity. The demonstration test of BAC proves that it can facilitate high quality of charging, while minimizing the electrical grid stress due to massive and concentrated charging of PHEVs and BEVs.

Author details

Muhammad Aziz

Address all correspondence to: aziz.m.aa@m.titech.ac.jp

Institute of Innovative Research, Tokyo Institute of Technology, Tokyo, Japan

References

[1] Aziz M, Oda T, Kashiwagi T. Extended utilization of electric vehicles and their re-used batteries to support the building energy management system. Energy Procedia. 2015;**75**:1938-1943. DOI: 10.1016/j.egypro.2015.07.226

[2] White CD, Max Zhang K. Using vehicle-to-grid technology for frequency regulation and peak-load reduction. Journal of Power Sources. 2011;**196**:3972-3980. DOI: 10.1016/j.jpowsour.2010.11.010

[3] Aziz M, Oda T, Mitani T, Watanabe Y, Kashiwagi T. Utilization of electric vehicles and their used batteries for peak-load shifting. Energies. 2015;**8**(5):3720-3738. DOI: 10.3390/en8053720

[4] Kempton W, Letendre S. Electric vehicles as a new power source for electric utilities. Transportation Research Part D: Transport and Environment. 1997;**2**:157-175. DOI: 10.1016/S1361-9209(97)00001-1

[5] Aziz M, Oda T, Morihara A, Murakami T, Momose N. Utilization of EVs and their used batteries in factory load leveling. In: Innovative Smart Grid Technologies Conference (ISGT), Washington: IEEE PES; 19-22 February 2014, 19 May 2014. pp. 1-5. DOI: 10.1109/ISGT.2014.6816370

[6] Donateo T, Ingrosso F, Licci F, Laforgia D. A method to estimate the environmental impact of an electric city car during six months of testing in an Italian city. Journal of Power Sources. 2014;**270**:487-498. DOI: 10.1016/j.jpowsour.2014.07.124

[7] Donateo T, Licci F, D'Elia A, Colangelo G, Laforgia D, Ciancarelli F. Evaluation of emissions of CO_2 and air pollutants from electric vehicles in Italian cities. Applied Energy. 2015;**157**:675-687. DOI: 10.1016/j.apenergy.2014.12.089

[8] Dyke KJ, Schofield N, Barnes M. The impact of transport electrification on electrical networks. IEEE Transactions on Industrial Electronics. 2010;**57**:3917-3926. DOI: 10.1109/TIE.2010.2040563

[9] Union of Concerned Scientists (UCS). BEV and PHEV Driving Range [Internet]. Available from: http://www.ucsusa.org/clean-vehicles/electric-vehicles/bev-phev-range-electric-car [Accessed: January 2017]

[10] Rahman I, Vasant PM, Singh BSM, Abdullah-Al-Wadud M, Adnan N. Review of recent trends in optimization techniques for plug-in hybrid, and electric vehicle charging infrastructures. Renewable and Sustainable Energy Reviews. 2016;**58**:1039-1047. DOI: 10.1016/j.rser.2015.12.353

[11] Jalilzadeh Hamidi R, Livani H. Myopic real-time decentralized charging management of plug-in hybrid electric vehicles. Electric Power Systems Research. 2017;**143**:522-532. DOI: 10.1016/j.epsr.2016.11.002

[12] Shaaban MF, Eajal AA, El-Saadany EF. Coordinated charging of plug-in hybrid electric vehicles in smart hybrid AC/DC distribution systems. Renewable Energy. 2015;**82**:92-99. DOI: 10.1016/j.renene.2014.08.012

[13] Nezamoddini N, Wang Y. Risk management and participation planning of electric vehicles in smart grids for demand response. Energy. 2016;**116**:836-850. DOI: 10.1016/j.energy.2016.10.002

[14] Aziz M, Oda T, Ito M. Battery-assisted charging system for simultaneous charging of electric vehicles. Energy. 2016;**100**:82-90. DOI: 10.1016/j.energy.2016.01.069

[15] Oda T, Aziz M, Mitani T, Watanabe Y, Kashiwagi T. Actual congestion and effect of charger addition in the quick charger station—Case study based on the records of expressway. IEEJ Transactions on Power and Energy. 2017;**136**(2):198-204. DOI: 10.1002/eej.22906

[16] Aziz M, Oda T. Load levelling utilizing electric vehicles and their used batteries. In: Fakhfakh MA, editor. Modeling and Simulation for Electric Vehicle Applications. InTech; 2016. pp. 125-147. DOI: 10.5772/64432

[17] Zhang L, Jabbari F, Brown T, Samuelsen S. Coordinating plug-in electric vehicle charging with electric grid: Valley filling and target load following. Journal of Power Sources. 2014:**267**;584-597. DOI: 10.1016/j.jpowsour.2014.04.078

[18] Jian L, Zhu X, Shao Z, Shao Z, Niu S, Chan CC. A scenario of vehicle-to-grid implementation and its double-layer optimal charging strategy for minimizing load variance within regional smart grids. Energy Conversion and Management. 2014;**78**:508-517. DOI: 10.1016/j.enconman.2013.11.007

[19] Nezamoddini N, Wang Y. Risk management and participation planning of electric vehicles in smart grids for demand response. Energy. 2016;**116**:836-850. DOI: 10.1016/j.energy.2016.10.002

[20] Khazali A, Kalantar M. A stochastic–probabilistic energy and reserve market clearing scheme for smart power systems with plug-in electrical vehicles. Energy Conversion and Management. 2015;**105**:1046-1058. DOI: 10.1016/j.enconman.2015.08.050

[21] CHAdeMO Association. High Power CHAdeMO [Internet]. Available from: https://www.chademo.com/technology/high-power/ [Accessed: 6 February 2017]

[22] CharIN e. V. Combined Charging System Specification [Internet]. Available from: http://www.charinev.org/ccs-at-a-glance/ccs-specification/ [Accessed: 6 February 2017]

[23] Tesla. Supercharger [Internet]. Available from: https://www.tesla.com/supercharger [Accessed: 6 February 2017]

[24] Yao JW, Wu F, Qiu XP, Li N, Su YF. Effect of CeO_2-coating on the electrochemical performances of $LiFePO_4$/C cathode material. Electrochimica Acta. 2011;**56**:5587-5592. DOI: 10.1016/j.electacta.2011.03.141

[25] Xiao LF, Cao YL, Ai XP, Yang HX. Optimization of EC-based multi-solvent electrolytes for low temperature applications of lithium-ion batteries. Electrochimica Acta. 2004;**49**:4857-4863. DOI: 10.1016/j.electacta.2004.05.038

[26] Ping P, Wang Q, Huang P, Sun J, Chen C. Thermal behaviour analysis of lithium-ion battery at elevated temperature using deconvolution method. Applied Energy. 2014;**129**:261-273. DOI: 10.1016/j.apenergy.2014.04.092

[27] Jansen AN, Dees DW, Abraham DP, Amine K, Henriksen GL. Low-temperature study of lithium-ion cells using a Li_ySn micro-reference electrode. Journal of Power Sources. 2007;**174**:373-379. DOI: 10.1016/j.jpowsour.2007.06.235

[28] Liao L, Zuo P, Ma Y, Chen X, An Y, Gao Y, Yin G. Effects of temperature on charge/discharge behaviors of $LiFePO_4$ cathode for Li-ion batteries. Electrochimica Acta. 2012;**60**: 269-273. DOI: 10.1016/j.electacta.2011.11.041

Chapter 4

A Hybrid Energy Storage System for a Coaxial Power-Split Hybrid Powertrain

Enhua Wang, Fuyuan Yang and Minggao Ouyang

Additional information is available at the end of the chapter

http://dx.doi.org/10.5772/67756

Abstract

A hybrid energy storage system (HESS) consisting of batteries and supercapacitors can be used to reduce battery stress and recover braking energy efficiently. In this paper, the performance of a novel coaxial power-split hybrid transit bus with an HESS is studied. The coaxial power-split hybrid powertrain consists of a diesel engine, a generator, a clutch, and a motor, whose axles are arranged in a line. A mathematical model of the coaxial power-split hybrid powertrain with an HESS is established and the parameters are configured using experimental data. Subsequently, to estimate the system performance, a program is designed based on Matlab and Advisor. A rule-based control strategy is designed and finely tuned for the coaxial power-split hybrid powertrain. Then, using the Chinese Transit Bus City Driving Cycle (CTBCDC), the system characteristics and energy efficiencies of the designed coaxial power-split hybrid powertrain with an HESS are analysed. The results indicate that the proposed coaxial power-split hybrid powertrain with an HESS can fulfil the drivability requirement of transit bus and enhance the energy efficiency significantly compared with a conventional powertrain bus as well as reduce the battery stress simultaneously. Using an HESS is a good solution for the coaxial power-split hybrid transit bus.

Keywords: coaxial power-split hybrid electric bus, hybrid energy storage system, supercapacitor, lithium-ion battery, performance simulation

1. Introduction

Energy conservation and emission reduction are two important tasks for a sustainable development of our industrial society. In terms of automotive industry, applications of hybrid electric

INTECH
open science | open minds

© 2017 The Author(s). Licensee InTech. This chapter is distributed under the terms of the Creative Commons Attribution License (http://creativecommons.org/licenses/by/3.0), which permits unrestricted use, distribution, and reproduction in any medium, provided the original work is properly cited.

vehicle (HEV), electric vehicle (EV), and fuel cell vehicle (FCV) are effective technical approaches for energy conservation and emission reduction [1, 2]. Currently, EV and HEV cannot comprehensively fulfill the requirements of transit buses due to their poor durability and high cost. However, HEV is a feasible solution with high reliability and relatively low cost.

There are two kinds of power source for an HEV: one is internal combustion engine, and the other is electric energy storage system (ESS). An ESS can discharge electric power to propel the vehicle or absorb electric power during the regenerative braking process. Generally, the architecture of an HEV powertrain can be classified as series hybrid, parallel hybrid, and power-split hybrid [3]. Taking into account the heavy-duty hybrid powertrain for transit buses, series hybrid and parallel hybrid are widely adopted presently. For instance, Orion VII and Man Lion use series hybrid powertrain for their transit buses. Volvo and Eaton designed two types of parallel hybrid powertrain for heavy-duty applications.

With the progress of technology, the battery used for the ESS of an HEV has been gradually shifted from lead-acid, NiCad, and Ni-MH to lithium-ion battery. Lithium-ion batteries will be used widely as the ESS for various vehicles because of their high-energy density, good safety, and long durability [4]. Currently, the materials of lithium-ion batteries are mainly lithium iron phosphate and nickel-cobalt-manganese ternary composite [5]. An HEV transit bus undergoes frequent acceleration and deceleration during its working time and requires large working currents of the ESS for these processes. Because the discharge and charge rates of lithium-ion battery are limited, if the ESS consists of only lithium-ion batteries, a large capacity of lithium-ion batteries is required, which will increase the cost and weight of an HEV greatly. To overcome this problem, hybrid energy storage system (HESS) composed of lithium-ion batteries and supercapacitors is employed. In contrast to a lithium-ion battery, a supercapacitor can charge or discharge with very large instantaneous currents. This characteristic can provide sufficient electric power during the acceleration process and store electric energy during the regenerative braking process. Because supercapacitors use a porous carbon-based electrode material, a very high effective surface area can be obtained by this porous structure compared to a conventional plate structure. Supercapacitors also have a minimal distance between the electrodes, which result in a very high capacitance compared to a conventional electrolytic capacitor [6, 7]. Apart from the fast charge/discharge rates and the high-power density, supercapacitors have much longer lifetimes (>100,000 cycles) compared to lithium-ion batteries [8–10]. However, supercapacitors normally have a much smaller energy capacity compared with lithium-ion batteries. Therefore, using an HESS can fully utilize the advantages of these two kinds of energy storage devices and avoid their disadvantages.

Figure 1 shows four primary topologies of an HESS, which encompass passive hybrid topology, supercapacitor semi-active hybrid topology, battery semi-active hybrid topology, and parallel active hybrid topology [11, 12]. The passive hybrid topology is the simplest to combine battery and supercapacitor together. The advantage of this topology is that no power electronic converters are needed. Because the voltage of the DC bus is stabilized by the battery, the stored energy of the supercapacitor cannot be utilized sufficiently. In the supercapacitor semi-active topology, the battery is connected to the DC bus directly, while the supercapacitor uses a bidirectional DC/DC converter to interface the DC bus. As a result, the voltage of the DC bus equals the output voltage of the battery so that it cannot be varied too much. But the voltage of the supercapacitor can be changed in a wide range. The disadvantage of this topology is that a large size of DC/DC converter

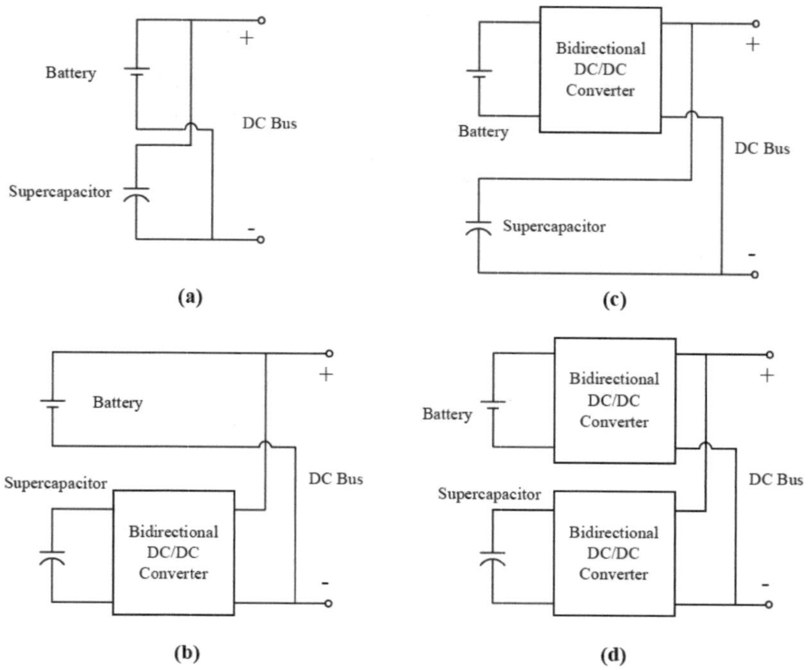

Figure 1. Topologies of hybrid energy storage system for vehicle application: (a) passive hybrid topology, (b) supercapacitor semi-active hybrid topology, (c) battery semi-active hybrid topology, and (d) parallel active hybrid topology.

is required. In the battery semi-active topology, the supercapacitor is connected to the DC bus directly, while the battery uses a bidirectional DC/DC converter to interface the DC bus. The advantage is the battery current can be maintained at a near constant value, thus the lifetime and energy efficiency of the battery can be improved significantly. The main shortcoming of this topology is that the voltage of the DC bus will vary during the charging/discharging process of the supercapacitor. The parallel hybrid topology is an optimal choice that can solve all the problems of the above topologies. The disadvantage is that two DC/DC converters are needed which will increase complexity, cost, and additional losses of the system. Three different topologies of HESS using supercapacitors as the main energy storage device for EV application were analyzed by Song et al. [13]. Rothgang et al. studied the performance of an active hybrid topology [14]. Among these various architectures, the supercapacitor semi-active hybrid topology using lithium-ion batteries as the main storage device is considered as a good solution for HEV application due to its high-power density and low cost.

Most of investigations about HESS focus on EV applications. Hung and Wu designed a rule-based control strategy for the HESS of an EV and estimated the system performance [15]. Vulturescu et al. tested the performance of an HESS consisting of NiCad batteries and supercapacitors [16]. Song et al. compared three different control strategies for an HESS, which included a rule-based control, a model predictive control, and a fuzzy logic control [17]. An HESS can also be used with a fuel cell and successfully satisfy the requirements of an FCV [18–20]. However, few investigations are concentrated on the application of an HESS for an HEV at present because they are more complicated than those for an EV. Masih-Tehrani et al. studied the energy management strategy of an HESS for a series hybrid powertrain [21] and Nguyen et al. investigated the performance of a belt-driven starter generator (BSG)-type parallel hybrid system with an HESS [22]. All these studies give the HESS an advantage over the ESS with only batteries.

The control strategy of an HESS for an EV is easier than that for an HEV or an FCV because it only needs to consider the power distribution between the batteries and supercapacitors. By contrast, the control strategy of an HEV with an HESS must assign the power demand of vehicle among the ICE, the batteries, and the supercapacitors, thus it is more complicated. Currently, few researches about the HESS for a power-split HEV are reported. In this study, a coaxial power-split HEV with an HESS for heavy-duty transit bus application is evaluated. The HESS uses lithium-ion batteries as the main energy storage device. The performance of the hybrid transit bus is analyzed using the Chinese Transit Bus City Driving Cycle (CTBCDC).

2. System description

Figure 2 shows the architecture of the designed coaxial power-split hybrid powertrain for a transit bus with a supercapacitor semi-active hybrid energy storage system. The auxiliary power unit (APU) consists of a diesel engine and a permanent magnetic synchronous generator (PMSG)

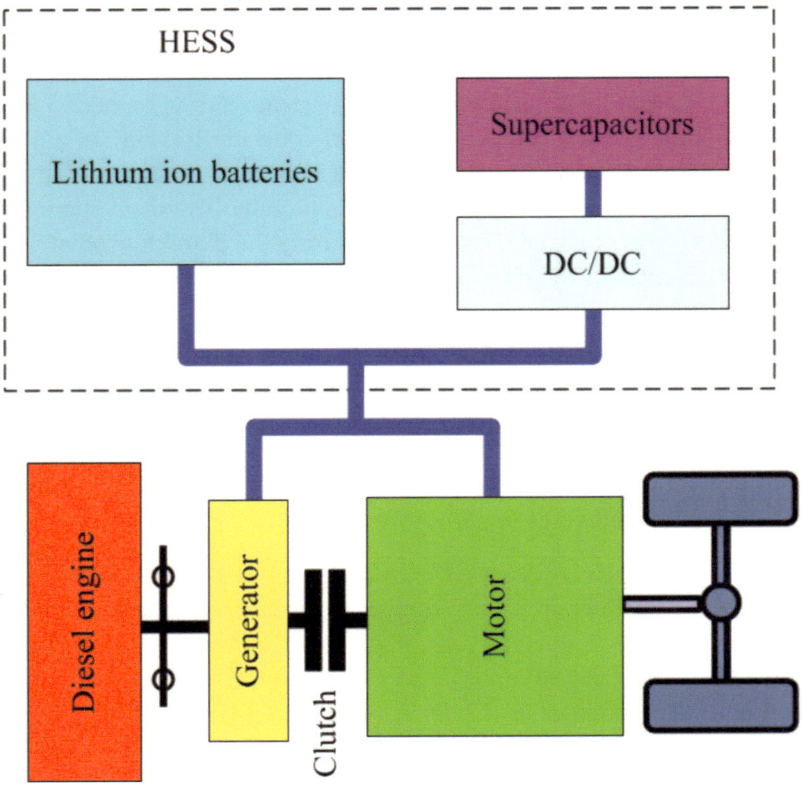

Figure 2. Architecture of the designed coaxial power-split hybrid powertrain with a hybrid energy storage system for transit bus where a CNG engine, a generator, a clutch, and a motor are arranged sequentially in a line.

http://dx.doi.org/10.5772/67756

whose shafts are connected directly. This shaft is also associated with the input axle of the clutch. The output axle of the clutch is linked to a permanent magnetic synchronous motor (PMSM) whose shaft is also connected to the final drive. Besides the diesel engine, an HESS composed of lithium-ion batteries and supercapacitors as well as a bidirectional DC/DC converter is used to provide electric power to the PMSM. A high-voltage power line is connected to the PMSG, the PMSM, the battery pack, and the DC/DC converter. The supercapacitor pack exchanges electric power with the high-voltage power line via the controllable bidirectional DC/DC converter.

The main technical parameters of the designed coaxial power-split hybrid powertrain with an HESS are listed in **Table 1**. The internal combustion engine is a 6.5 L YC6J220 diesel engine manufactured by Yuchai Machinery Co., Ltd. The supercapacitor pack is composed serially of 12 units of Maxwell 48-V module. The battery pack consists of nine paralleled groups and each

Component	Parameter	Value	Unit
Engine	Type	YC6J220 diesel engine	–
	Rated power	162	kW
	Rated speed	2500	r/min
	Maximum torque	760	N·m
	Peak efficiency	41.9	%
Supercapacitors	Type	Maxwell 48-V module	–
	Number of modules	12	–
	Energy capacity	0.651	kWh
	Rated voltage	576	V
	Voltage range	172–570	V
Batteries	Type	A123 26650 $LiFePO_4$ cell	–
	Number of modules	1701	–
	Energy capacity	14	kWh
	Rated voltage	604	V
	Voltage range	570–591	V
Motor	Type	Jing-Jin Electric PMSM	–
	Rated power	150	kW
	Rated speed	3000	r/min
	Maximum torque	2100	N·m
	Peak efficiency	93.5	%
Generator	Type	Jing-Jin Electric PMSG	–
	Rated power	135	kW
	Rated speed	3000	r/min
	Maximum torque	850	N·m
	Peak efficiency	94.5	%

Table 1. Technical specifications of the coaxial power-split hybrid powertrain.

group is serially connected by 189 cells of the A123 nanophosphate lithium-ion ANR26650M1-B cell. The energy capacity of the battery pack is decided by the all-electric driving range, and its c-rate is defined by the average output power. In this design, the energy capacity of the battery pack is 14 kWh. Both the PMSG and the PMSM are low-speed high-power electric machines developed by Jing-Jin Electric Technologies Co., Ltd.

3. Mathematical model

A mathematical model of the coaxial power-split hybrid powertrain with an HESS is established to analyze the system performance. According to the working principle of the entire system, a lumped-parameter model is used.

3.1. Vehicle dynamics

Since the rear axle is used to drive the hybrid transit bus, the tractive force of the rear axle is determined according to the longitudinal vehicle dynamics [23].

$$\mathrm{F} = \mathrm{F}_r + \mathrm{F}_a + \mathrm{F}_i + \mathrm{F}_j = mg\cos\alpha(f_1 + f_2 v) + \frac{1}{2}\rho A C_D v_r^2 + mg\sin\alpha + \delta m\frac{dv}{dt} \tag{1}$$

Therefore, the tractive force F can be calculated according to the practical driving conditions.

The loads of the front and rear axles can be calculated based on the vehicle parameters. Then, the friction factor and the slip of the tires can be determined due to the required tractive force and the axle load. Moreover, the rotating speed and the driving torque of the tires can be obtained. If the vehicle is braking, the braking force imposed on the rear axle can be computed according to the braking force distribution equation. Then, the braking torque and speed of the rear tires can be determined. Subsequently, the model of the final drive determines the rotating speed ω_m and the torque T_m taking into account the friction loss T_{l0} and the rotating inertia J_0.

$$\mathrm{T}_m = \frac{T_w}{i_0} + T_{l0} + i_0 J_0 \frac{d\omega}{dt} \tag{2}$$

$$\omega_m = i_0\omega \tag{3}$$

The PMSM model accounts for the requested torque T_{mr} and the input power P_m calculated by a two-dimensional (2D) lookup table measured from a motor test bench [24].

In the coaxial power-split hybrid powertrain, a clutch is used to control the operation mode that is either the series or the parallel mode. The mathematical model of the clutch is a simple friction model [25], which is used to determine the clutch state among the engaged, the slipping, and the disengaged states. The torque and speed transmitted through the clutch are also determined.

http://dx.doi.org/10.5772/67756

3.2. Auxiliary power unit

Normally, an auxiliary power unit is a combination of the engine and the generator. The diesel engine model accounts for the requested torque T_{er} according to the engine output torque T_e and the engine speed ω_e.

$$\mathrm{T}_{er} = \mathrm{T}_e + \mathrm{J}_e \frac{d\omega_e}{dt} \tag{4}$$

Subsequently, the fuel consumption M_e is calculated using the density of diesel fuel ρ_f for the diesel engine.

$$\mathrm{M}_e = \frac{\int_0^{t_c} m_e dt}{\rho_f \int_0^{t_c} v dt} \tag{5}$$

where t_c is the final time of the driving cycle. The mathematical model of the generator is similar with the motor. The requested torque T_{gr} is determined according to the input torque of the generator T_g and the inertia J_g. The output power P_g is a function of the generator speed ω_g and torque T_{gr} defined by a 2D lookup table.

$$\mathrm{T}_{gr} = \mathrm{T}_g + \mathrm{J}_g \frac{d\omega_g}{dt} \tag{6}$$

$$\mathrm{P}_g = f(\omega_g, T_{gr}) \tag{7}$$

3.3. Hybrid energy storage system

The model of lithium-ion batteries of the designed HESS is a *Rint* model [24]. If the output power of the battery pack is P_{bat}, the energy equilibrium of this circuit can be expressed as

$$\mathrm{R}_{\text{bat}} i_{\text{bat}}^2 - U_{\text{bat}} i_{\text{bat}} + P_{\text{bat}} = 0 \tag{8}$$

Then, the current of the battery is calculated by

$$i_{\text{bat}} = \frac{U_{\text{bat}} - \sqrt{U_{\text{bat}}^2 - 4R_{\text{bat}}P_{\text{bat}}}}{2R_{\text{bat}}} \tag{9}$$

The *SOC* of the lithium-ion batteries at time *k* is defined as

$$SOC_{\text{bat},k} = SOC_{\text{bat},0} + \int_{t=0}^{k} i_{\text{bat}}(t) dt \tag{10}$$

The internal resistance R_{bat} and capacitance C_{bat} are calculated via looking up two 2D maps. The data of these maps are obtained according to the references [26, 27] for A123 system's ANR26650M1-B lithium-ion battery.

The supercapacitors of the designed HESS use an *RC* model [24]. Given the output power of the supercapacitors P_{sc}, the energy equation of the supercapacitor circuit is expressed as

$$R_{sc}i_{sc}^2 - U_{sc}i_{sc} + P_{sc} = 0 \tag{11}$$

Then, the current of the supercapacitors is obtained by

$$i_{sc} = \frac{U_{sc} - \sqrt{U_{sc}^2 - 4R_{sc}P_{sc}}}{2R_{sc}} \tag{12}$$

The SOC of the supercapacitors at time k is calculated as

$$SOC_{sc}(k) = \frac{U_{sc}(k-1) - \frac{i_{sc}(k-1)}{C_{sc}(k-1)}}{U_{sc,\max}} \tag{13}$$

To make sure the supercapacitors can recover the regenerative braking energy efficiently, the SOC of the supercapacitors is corrected according to the vehicle velocity during the driving process.

$$SOC_{sc}^*(k) = SOC_{sc,\max}\sqrt{1 - \left(1 - \frac{U_{sc,\min}^2}{U_{sc,\max}^2}\right)\left(\frac{v(k)}{v_{\max}}\right)^2} \tag{14}$$

where SOC^*sc is the corrected SOC, $SOC_{sc,max}$ is the maximum SOC, $U_{sc,max}$ and $U_{sc,min}$ are the maximum and minimum voltages of the supercapacitors, respectively, and v_{max} is the maximum vehicle velocity. The designed supercapacitor pack consists of Maxwell's 48-V module and the detailed parameters are obtained from a laboratory test.

The DC/DC model of the HESS uses a 2D map to determine the energy efficiency based on the input power and the voltage ratio, which can provide a good precision and a fast calculation speed.

4. Energy management strategy design

The control strategy of a coaxial power-split hybrid powertrain only using supercapacitors as energy storage system was designed by Ouyang et al. previously, which involves series hybrid mode, parallel hybrid mode, and the mode transition logic [28]. For the coaxial power-split hybrid powertrain with an HESS, one more task must be designed—the energy management strategy of the HESS. The thermostatic control, the power follower control, and the optimal control [29] are the three main control strategies for the series hybrid mode while the parallel electric assist control, the adaptive control, and the fuzzy logic control are normally used for the parallel hybrid mode. Currently, the rule-based control, the filter control, the model predictive control, and the fuzzy logic control are the four main control strategies of an HESS to distribute the power demand between the batteries and the supercapacitors.

Investigations indicate that the performance of rule-based control strategy can approach to that of the optimal control after optimization of the parameters of the rule-based control [30].

http://dx.doi.org/10.5772/67756

Therefore, in this study, a rule-based control strategy for the coaxial power-split hybrid powertrain is designed and is shown in **Figure 3**. The entire rule-based control strategy is composed of a series mode control and a parallel mode control as well as an HESS control. The series mode control strategy uses a power follower control method shown in **Figure 3**. If the driving power demand of the vehicle is lower than a certain value, the system enters the series mode control. First, the required power of the motor P_d is computed according to the system mathematical model. Then, the output power of the engine is determined based on the generator efficiency map. The modified discharging/charging power is calculated according to the difference between the present SOC and the target one. Subsequently, the requested power of the diesel engine is determined according to the defined optimal operation line (OOL) of the series mode control [31]. Meanwhile, the engine on/off state is decided due to the required power P_d and the current SOC. If the engine state is on, the final required engine torque and speed are computed according to the requested power of the engine and the defined OOL.

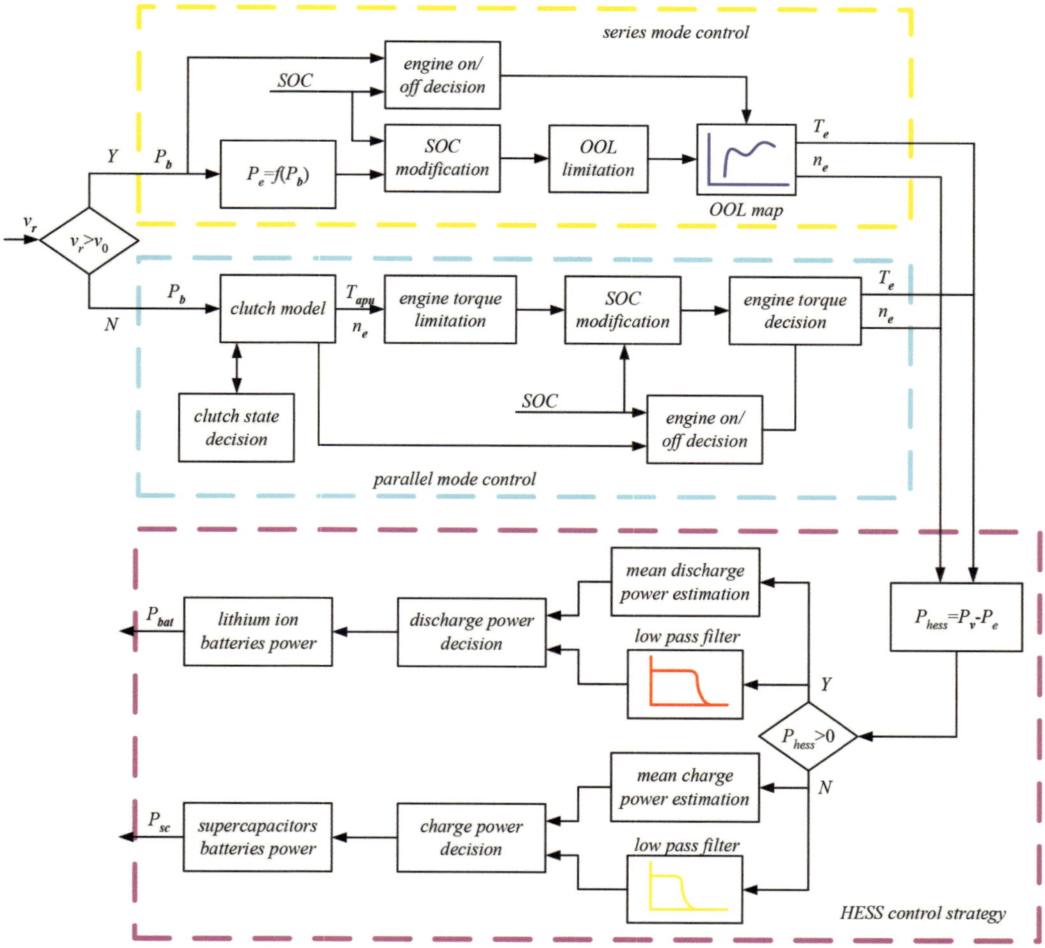

Figure 3. Control strategy for the coaxial power-split hybrid powertrain with an HESS composed of a series mode control and a parallel mode control as well as an HESS control strategy.

If the vehicle velocity is greater than the set value, the system changes to the parallel mode control. A parallel electric assist control strategy is used in this research and is shown in **Figure 3**. First, the demanded speed and torque of the auxiliary power unit are determined by the clutch model according to the power demand of the motor P_d. In the clutch model, the decision logic of the clutch state sets the working state of the clutch. When the clutch is engaged at the parallel mode control, the requested torque of the APU, which equals the output torque of the engine, is the summation of the driving torque directly transferred to the final drive and the charge torque that is transmitted to the generator. The motor provides the remaining driving torque simultaneously. Subsequently, the control parameters of the parallel mode are optimized across the overall engine's working region. Same as the series mode control, the modified discharging/charging power is calculated. Then, engine on/off state is decided by the clutch state and the SOC as well as the requested vehicle velocity. At last, the requested torque and speed of the engine are determined based on the engine state and the power demand.

The requested power of the diesel engine is obtained by the series or parallel control strategy. Then, the power demand of the HESS P_{hess} can be calculated as the difference between the power demand of the motor and the request power of the diesel engine. The HESS energy management strategy is in charge of the distribution of the power demand between the lithium-ion batteries and the supercapacitors. The detailed strategy shown in **Figure 3** is a combination of the rule-based control and the filter control. If P_{hess} is positive, the HESS should output electric power to the power line. First, an algorithm estimates the mean discharge power during the driving cycle and a low-pass filter outputs a filtered discharge power of the batteries. Then, the discharging power decision block calculates the output powers of the batteries and the supercapacitors based on the designed threshold values of SOC and the mean discharge power. If P_{hess} is negative, the motor outputs electric power to charge the HESS. Thereby, the control strategy is similar with the positive situation. Meanwhile, an SOC correction algorithm of the supercapacitors is employed based on the vehicle velocity for the discharging process.

5. Results and discussion

A program was developed using Matlab and Advisor according to the mathematical model of the coaxial power-split hybrid powertrain with an HESS. The designed simulation model combines the backward- and forward-facing methods and can evaluate the drivability and economy of the coaxial power-split hybrid powertrain system. In this research, a hybrid transit bus with a total mass of 15 ton manufactured by Higer Bus Company Limited is applied. The detailed parameters of the hybrid transit bus are listed in **Table 2**.

Using the CTBCDC driving cycle, the performance of the designed coaxial power-split hybrid powertrain with an HESS is evaluated, and the results are shown in **Figure 4**. The target and available values of the vehicle speed are compared in **Figure 4a**. The target vehicle speed is defined by the CTBCDC driving cycle. The available vehicle speed is determined using the established program and can follow the target one perfectly, which means the coaxial power-split hybrid powertrain with an HESS can satisfy the requirements of the drivability completely. The evaluated power demand of the motor is shown as the solid profile of **Figure 4b**. As a contrast, the output power of the diesel engine is given as the dashed line.

Parameter	Value	Unit
Vehicle mass excluding pack	12,930	kg
Cargo mass	2070	kg
Dimensions	12 × 2.55 × 3	m × m × m
Rolling resistance coefficient	0.0094	–
Aerodynamic drag coefficient	0.70	–
Vehicle frontal area	7.65	m^2
Wheel radius	0.506	m
Final gear	5.833	–

Table 2. Parameters of the transit bus.

The designed coaxial power-split hybrid powertrain with an HESS can recover the braking energy efficiently during the regenerative braking. When the vehicle is driving, the power demand of the motor increases with the vehicle velocity and the diesel engine will start and provide part of the driving energy if the driving power demand of the vehicle is greater than the specified value of the series mode control or the parallel mode control.

The system operation state is shown in **Figure 4c**, where 0 represents the EV mode, 1 means the series control mode, and 2 denotes the parallel control mode. If the vehicle driving power is greater than 110 kW, the operation mode switches from the series mode to the parallel mode. Meanwhile, the EV mode can be changed from both the series control mode and the parallel control mode. The output speed and torque of the diesel engine are shown in **Figure 4d** and **e**, respectively. Because the power follower control is used for the series control mode and the engine speed is proportional to the vehicle velocity at the parallel mode, the engine speed varied from 1494 to 1842 r/min during the working process. Furthermore, the engine output torque maintains at a high level of 550.4–761.6 Nm when the engine is running.

The performance of the HESS for the coaxial power-split hybrid powertrain is displayed in **Figure 5**. The output power of the HESS is given in **Figure 5a**. The output powers of each of the energy storage devices are given in **Figure 5b**, where the dashed line is used for the lithium-ion batteries, and the dash dotted line is used for the supercapacitors. In **Figure 5**, positive values represent discharging process and negative is used for charging. The output power of the HESS varies with the power demand of the motor. The maximum and minimum output powers achieve 104.8 and −112.9 kW, respectively. The output power of the lithium-ion batteries changes from −50 to 50 kW during the driving cycle, which means the output current of the batteries can be significantly decreased. The supercapacitors discharge or charge with large powers for the high-power working conditions to compensate the power difference between P_{hess} and P_{bat}. The SOC profiles for the lithium-ion batteries and the supercapacitors are shown in **Figure 6c**. During the charge sustain mode, the SOC of the batteries manifests a very small variation during the cycle. As a contrast, the SOC of the supercapacitors varies within a wide range from 0.395 to 0.99. The reason is that the capacity of the lithium-ion batteries is much greater than that of the supercapacitors and the output power of the lithium-ion batteries is constrained to a small range. The output voltages of the batteries and the supercapacitors are shown in **Figure 5d**. The output voltage of the lithium-ion batteries, which equals the power line voltage, changes in a very small

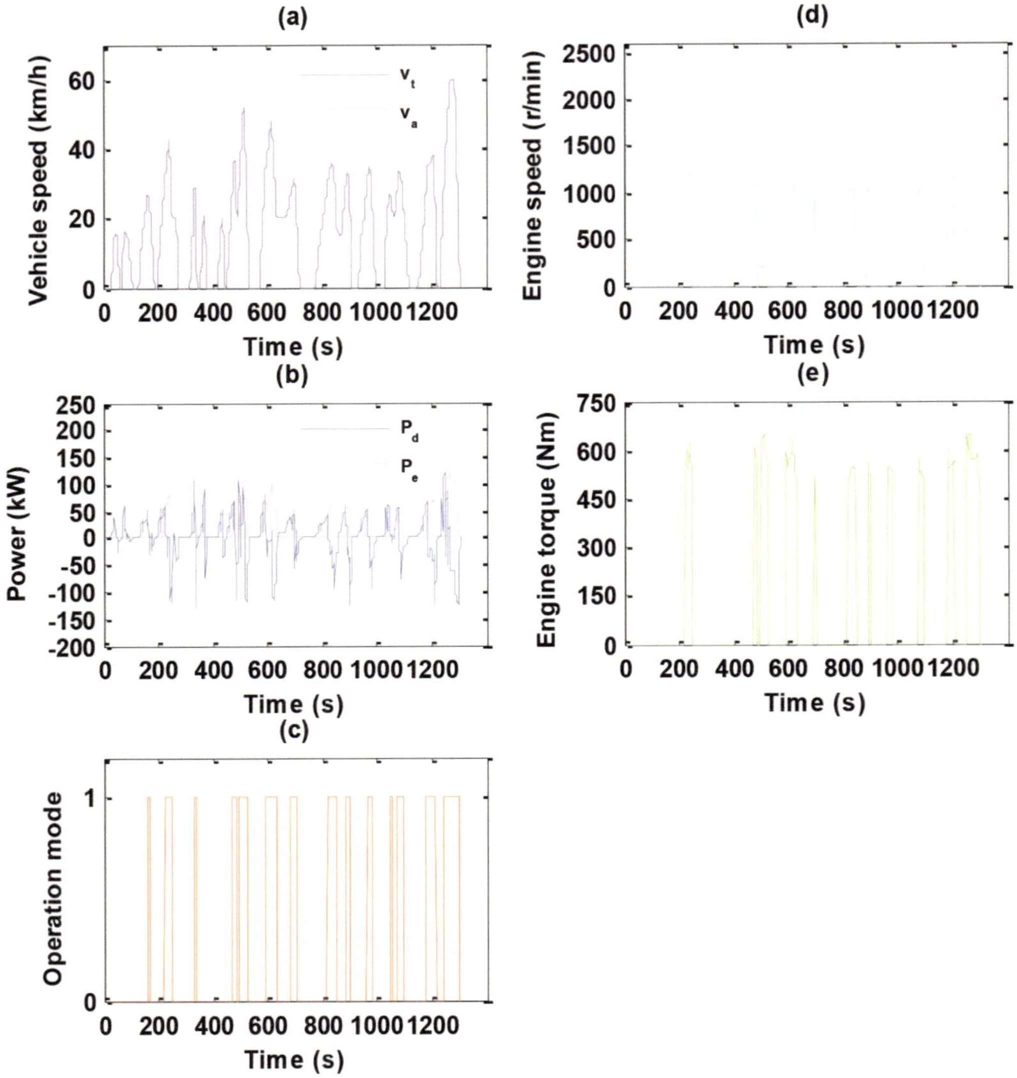

Figure 4. Analysis results of the system performance of the designed coaxial power-split hybrid powertrain with an HESS using the Chinese Transit Bus City Driving Cycle: (a) comparison of the target and estimated vehicle speeds, (b) the driving power requested by the transit bus P_d and the requested power provided by the engine P_e, (c) the system operation mode where 0 means pure electric vehicle mode and 1 represents series hybrid mode and 2 is parallel hybrid mode, (d) the engine speed, and (e) the engine requested torque.

range from 559.1 to 584.9 V. The stable voltage characteristic will be beneficial for the operations of the motor and the generator. Although the output voltage of the supercapacitors varies in a large scope, it is still within the allowable voltage ratio of the DC/DC module. The currents of the batteries and the supercapacitors are displayed in **Figure 5e**. The current of the supercapacitors increases while the current of the batteries is much smaller and the discharging/charging rate of the batteries is less than 4C, which is very helpful for the life extension of the batteries.

http://dx.doi.org/10.5772/67756

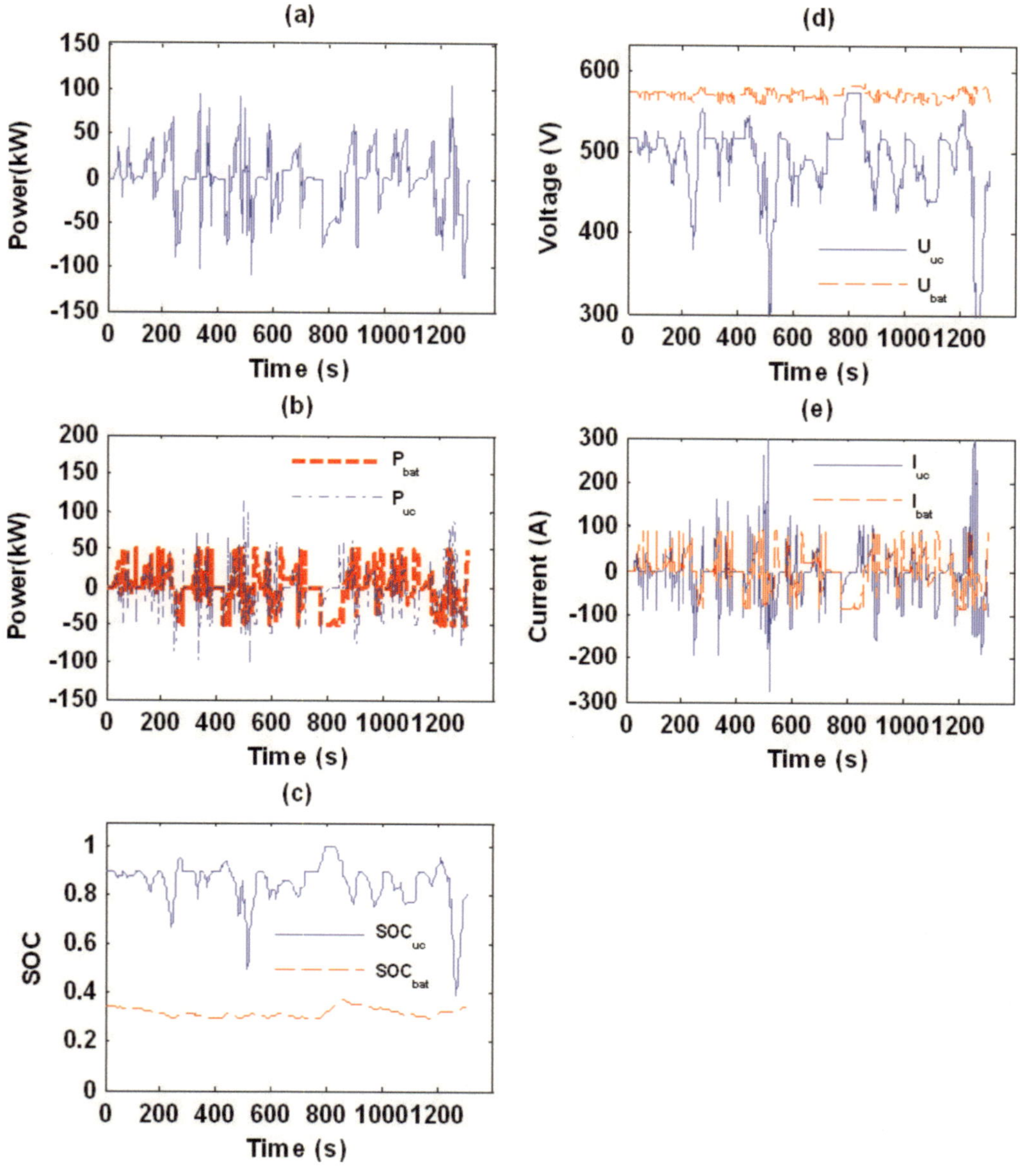

Figure 5. Analysis results for the performance of the HESS system: (a) the electric power output by the HESS, (b) the power outputs of the batteries P_{bat} and the supercapacitors P_{sc}, (c) SOCs of the batteries SOC_{bat} and the supercapacitors SOC_{sc}, (d) output voltages of the batteries U_{bat} and the supercapacitors U_{sc}, (e) output currents of the batteries i_{bat} and the supercapacitors i_{sc} where positive values denote discharging process and negative for charging.

Figure 6 gives the performances of the motor and the generator of the coaxial power-split hybrid powertrain with an HESS. The input power of the motor generally follows the vehicle driving power during the cycle, which means only a small part of the mechanical power of the diesel engine is used to propel the vehicle directly. Although the current of the motor varies in a wide range from −193.6 to 215.9 A, it does not exceed the maximum operation current of the motor.

The generator operates only a small part of the time and its output power changes mainly from 24.38 to 104.8 kW. The current of the generator varies with the output power because the power line voltage is stable.

To evaluate the potential of fuel saving for the coaxial power-split hybrid powertrain with an HESS, 12 prototypes of hybrid transit bus were built and applied to a practical city routes in Ningbo, Zhejiang Province. A total mileage of over 40,000 km for each hybrid transit bus was achieved, and the average fuel consumption is approximately 24.53 L/100 km, which is listed in **Table 3**. Using the established mathematical model and the analysis program, the estimated fuel consumption of the coaxial power-split hybrid powertrain with an HESS is 24.43 L/100 km, which is also listed in **Table 3**. The SOC difference of the lithium-ion batteries between the start and end points is 0.0034. The practical driving routes in Ningbo are different with the CTBCDC driving cycle. The start/stop frequency is decreased compared with the CTBCDC driving cycle. Therefore, the improvement of fuel efficiency is a bit lower. On the other hand, the ambient temperature and the total vehicle weight varying during the practical driving conditions also

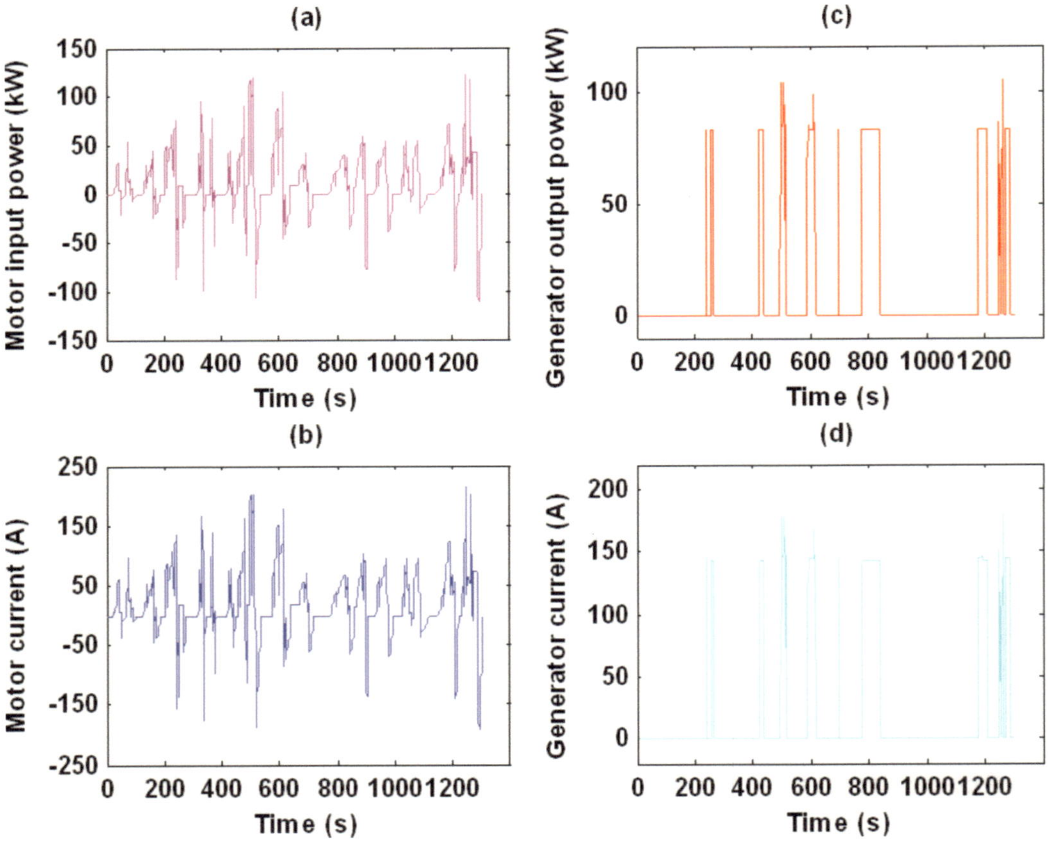

Figure 6. Results of the motor and the generator: (a) the input power of the motor (a positive value is used for driving process and a negative one is for regenerative braking), (b) the output current of the motor, (c) the requested power of the generator, and (d) the output current of the generator.

have a great influence on the fuel consumption. Herein, the experimental result only gives a very coarse and average evaluation of the fuel efficiency. Moreover, the fuel consumption of a conventional powertrain using a YC6G270 diesel engine and a five-gear manual transmission is computed based on the same vehicle parameters as the hybrid transit bus. The total mass of the bus with the conventional powertrain is 15,000 kg, which is the same as that of the coaxial power-split hybrid powertrain with an HESS. The result is 36.33 L/100 km. Compared with the results of the conventional powertrain, the fuel consumption of the coaxial power-split hybrid powertrain with an HESS can be decreased significantly by about 32.5%.

From the viewpoint of energy efficiency, the reason for such a great fuel reduction can be explained. **Figure 7** shows the effective thermal efficiency map of the YC6J220 diesel engine obtained on an engine test bench. The engine's working points estimated by the analysis program are also displayed. The OOL line of the series control mode is represented by the thick solid line in **Figure 7**. As a contrast, the effective thermal efficiency map of the YC6G270 diesel engine for the conventional powertrain and the corresponding working points are given in **Figure 8**. It can be seen that the engine working points of the coaxial power-split hybrid powertrain with an HESS are very close to the region having the peak efficiency and their thermal efficiencies are greater than 40%. However, the working points of the conventional powertrain shown in **Figure 8** will change with the vehicle velocity, resulting in a very wide distribution from the idle speed to the full load. Therefore, many working points of the conventional powertrain locate in the low-efficiency regions, leading to a low efficiency of the entire powertrain system.

The energy efficiency map of the PMSM obtained on a motor test bench and the relative working points for the CTBCDC driving cycle are given in **Figure 9**. A large part of the working points situates close to the peak-efficiency region. The efficiencies of most of the working points are higher than 93.4% except for the low-speed and small-load regions. The energy-weighted average efficiency of the motor during the CTBCDC driving cycle is 91.92%. Because the motor is connected to the final drive without a transmission, a particular design of the PMSM can ensure that the motor efficiency is high enough for low-speed working conditions. The energy efficiency map of the PMSG measured on a motor test bench and the corresponding working points for the CTBCDC driving cycle are displayed in **Figure 10**. The efficiencies of the PMSG are found to be between 92 and 93% during the CTBCDC driving cycle, and the energy-weighted average efficiency is 92.55%, which approaches the peak efficiency of the PMSG

Powertrain	Fuel consumption (l/100 km)	Energy reduction[a] (%)
Conventional powertrain bus	36.33	
Coaxial power-split hybrid bus	24.43[b]	32.76
	24.53[c]	32.48

[a]Relative to the conventional bus equipped with a YC6G270 diesel engine.
[b]Analysis result of the hybrid transit bus equipped with a YC6J220 diesel engine.
[c]Test result of the hybrid transit bus equipped with a YC6J220 diesel engine.

Table 3. Results of fuel consumption.

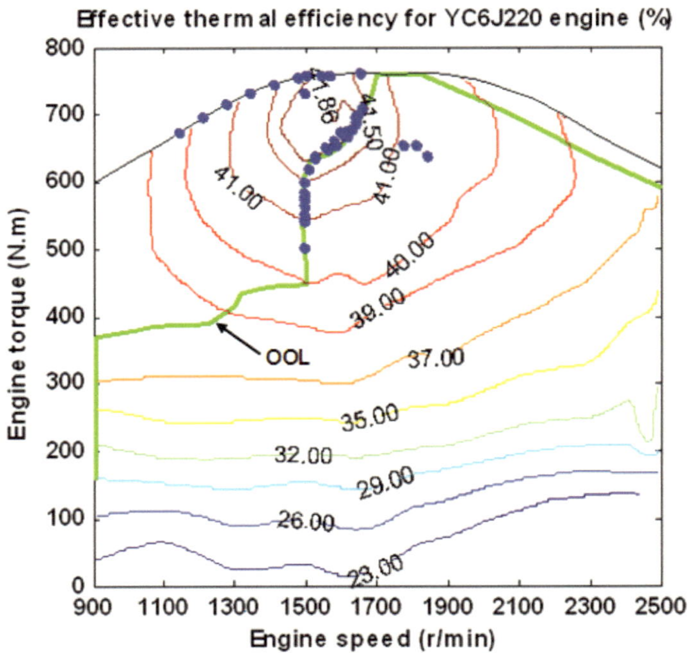

Figure 7. Energy efficiency map of the YC6J220 diesel engine measured on a test bench where the points denote the operation conditions of the diesel engine for the coaxial power-split hybrid powertrain with an HESS.

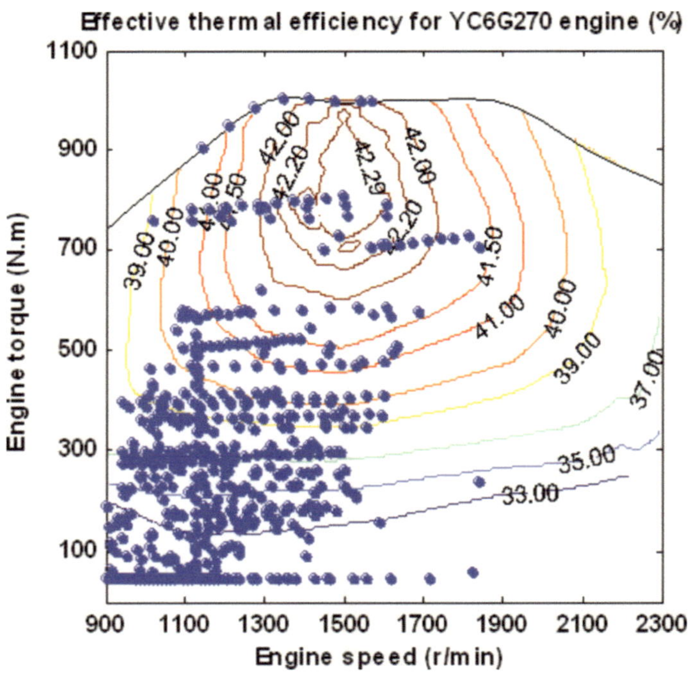

Figure 8. Analysis results of the conventional powertrain using a YC6G270 diesel engine and a five-gear transmission.

http://dx.doi.org/10.5772/67756

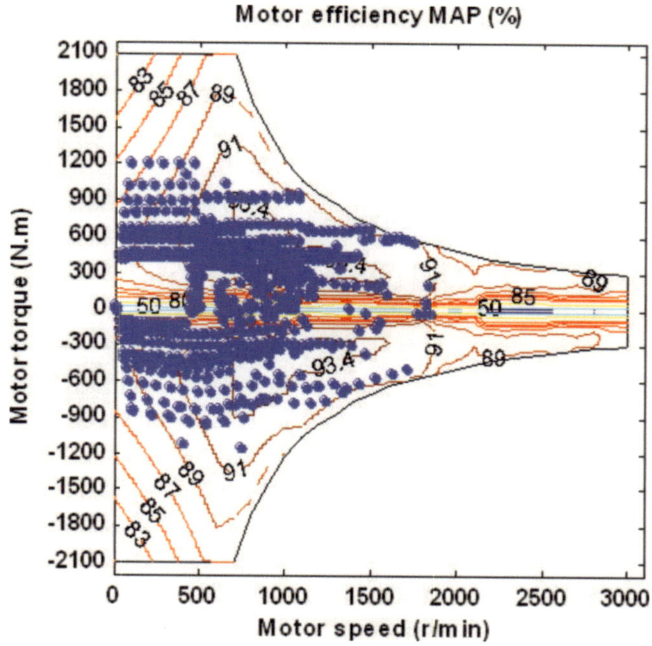

Figure 9. Energy efficiency map of the motor measured on a motor test bench where the points denote operation points of the motor for the coaxial power-split hybrid powertrain with an HESS.

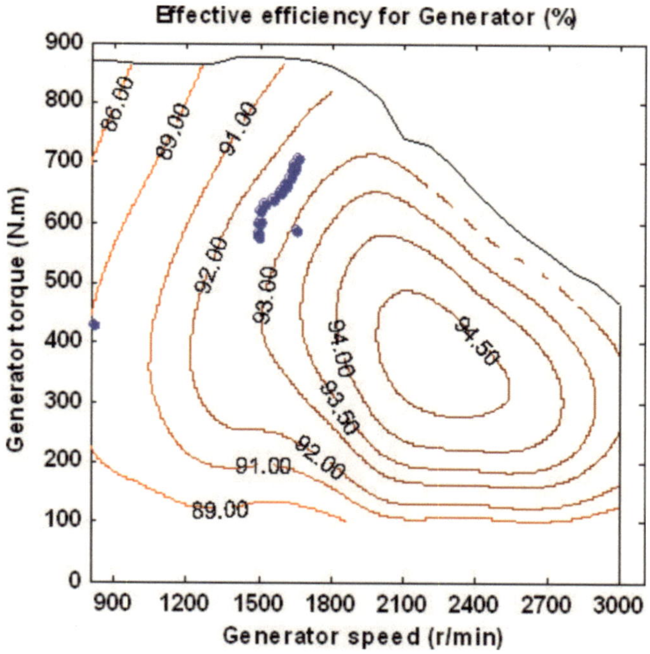

Figure 10. Energy efficiency map of the generator measured on a motor test bench where the points denote operation points of the generator for the coaxial power-split hybrid powertrain with an HESS.

6. Conclusions

In this study, the system performance of a coaxial power-split hybrid powertrain with an HESS for transit bus application was investigated. First, a system topology was designed. Then, a mathematical model was established and an energy management strategy was developed. Finally, the energy efficiency of the hybrid powertrain system was evaluated by Matlab and Advisor. The main conclusions are summarized as follows:

1. Compared with the results of a coaxial power-split hybrid powertrain with supercapacitors as energy storage device, the equivalent fuel consumption of the designed coaxial power-split hybrid powertrain with an HESS is a bit higher (24.43 vs. 20.46 L/100 km). The reason is that the energy capacity of the supercapacitors in this study is much smaller than that of the hybrid powertrain with only supercapacitors. Thus, the amount of the recovered regenerative energy is smaller than that of the hybrid powertrain with only supercapacitors. On the other hand, the DC/DC converter and the battery pack also have some losses during the working processes. As a result, the energy efficiency of the coaxial power-split hybrid powertrain with an HESS is lower than that of the hybrid powertrain with supercapacitors. Because we have no data about the coaxial power-split hybrid powertrain with only batteries, a quantitative comparison cannot be given here. Generally, if only batteries are used for the energy storage system, a large size will be used and the cost will increase a lot. The energy efficiency of such a system will be a trade-off between cost and lifetime of the batteries. Nevertheless, the results of this study show that the coaxial power-split hybrid powertrain with an HESS has a very good energy efficiency compared with a conventional powertrain of the same level.
2. Because the designed HESS has a smaller size and weight and lower price than that with only supercapacitors, and the lithium-ion batteries of the HESS can operate at an averaged current and thus have a longer life cycle, the HESS is more suitable for hybrid transit bus application.
3. The power line voltage of the HESS is more stable than that with only supercapacitors. This will be beneficial for the operation of the accessories such as the air conditioner or the in-vehicle infotainment system during the driving process.

Nomenclature

A vehicle frontal area (m^2)
C capacitance (F)
C_D aerodynamic drag coefficient
F tractive force (N)
f rolling resistance coefficient
g standard gravitational acceleration (9.8067 m/s^2)
i current (A)
J inertia ($kg \cdot m^2$)

http://dx.doi.org/10.5772/67756

M	fuel consumption (L/100 km)
m	vehicle mass (kg)
P	power (W)
R	resistance (Ω)
T	torque (Nm)
t	time (s)
U	voltage (V)
v	vehicle velocity (m/s)

Greek letters

α	road angle (°)
δ	mass factor
ρ	air density (kg/m^3)
ω	angular speed of a tire

Subscript

0	final drive
a	aerodynamic
b	braking
bat	battery
e	engine
f	fuel
g	generator
i	grade resistance
j	acceleration
l	loss
m	motor
r	rolling, rear, reduced
sc	supercapacitor
w	wheel

Acronyms

APU	auxiliary power unit
CTBCDC	Chinese Transit Bus City Driving Cycle
ESS	energy storage system
EV	electric vehicle
FCV	fuel cell vehicle
HESS	hybrid energy storage system

HEV	hybrid electric vehicle
ICE	internal combustion engine
OOL	optimal operation line
PMSG	permanent magnetic synchronous generator
PMSM	permanent magnetic synchronous motor
SOC	state of charge

Author details

Enhua Wang[1], Fuyuan Yang[2] and Minggao Ouyang[2*]

*Address all correspondence to: ouymg@tsinghua.edu.cn

1 School of Mechanical Engineering, Beijing Institute of Technology, Beijing, China

2 State Key Laboratory of Automotive Safety and Energy, Tsinghua University, Beijing, China

References

[1] Kuhne R. Electric buses—An energy efficient urban transportation means. Energy. 2010;**35**:4510–4513.

[2] Chiara F, Canova M. A review of energy consumption, management, and recovery in automotive systems, with considerations of future trends. Proc I Mech Eng D-J Aut. 2013;**227**:914–936.

[3] Lajunen A. Energy consumption and cost-benefit analysis of hybrid and electric city buses. Transport Res C. 2014;**38**:1–15.

[4] Shen J, Dusmez S, Khaligh A. Optimization of sizing and battery cycle life in battery/ultracapacitor hybrid energy storage systems for electric vehicle applications. IEEE T Ind Electron. 2014;**10**:2112–2121.

[5] Capasso C, Veneri O. Experimental analysis on the performance of lithium based batteries for road full electric and hybrid vehicles. Appl Energ. 2014;**136**:921–930.

[6] Koohi-Kamali S, Tyagi VV, Rahim NA, Panwar NL, Mokhlis H. Emergence of energy storage technologies as the solution for reliable operation of smart power systems: A review. Renew Sust Energ Rev. 2013;**25**:135–165.

[7] Vangari M, Pryor T, Jiang L. Supercapacitors: Review of materials and fabrication methods. J Energy Eng. 2013;**139**:72–79.

[8] Zhang LL, Zhao XS. Carbon-based materials as supercapacitor electrodes. Chem Soc Rev. 2009;**38**:2520–2531.

http://dx.doi.org/10.5772/67756

[9] Yu Z, Duong B, Abbitt D, Thomas J. Highly ordered MnO_2 nanopillars for enhanced supercapacitor performance. Adv Mater. 2013;**25**:3302–3306.

[10] Gao Y, Zhou YS, Qian M, He XN, Redepenning J, Goodman P, Li HM, et al. Chemical activation of carbon nano-onions for high-rate supercapacitor electrodes. Carbon. 2013;**51**:52–58.

[11] Kuperman A, Aharon I. Battery–ultracapacitor hybrids for pulsed current loads: A review. Renew Sust Energ Rev. 2011;**15**:981–992.

[12] Cao J, Emadi A. A new battery/ultracapacitor hybrid energy storage system for electric, hybrid, and plug-in hybrid electric vehicles. IEEE T Power Electr. 2012;**27**:122–132.

[13] Song Z, Hofmann H, Li J, Han X, Zhang X, Ouyang M. A comparison study of different semi-active hybrid energy storage system topologies for electric vehicles. J Power Sources. 2015;**274**:400–411.

[14] Rothgang S, Baumhöfer T, van Hoek H, Lange T, De Doncker RW, Sauer DU. Modular battery design for reliable, flexible and multi-technology energy storage systems. Appl Energ. 2015;**137**:931–937.

[15] Hung YH, Wu CH. An integrated optimization approach for a hybrid energy system in electric vehicles. Appl Energ. 2012;**98**:479–490.

[16] Vulturescu B, Trigui R, Lallemand R, Coquery G. Implementation and test of a hybrid storage system on an electric urban bus. Transport Res C. 2013;**30**:55–66.

[17] Song Z, Hofmann H, Li J, Hou J, Han X, Ouyang M. Energy management strategies comparison for electric vehicles with hybrid energy storage system. Appl Energ. 2014;**134**:321–331.

[18] Yu Z, Zinger D, Bose A. An innovative optimal power allocation strategy for fuel cell, battery and supercapacitor hybrid electric vehicle. J Power Sources. 2011;**196**:2351–2359.

[19] Hu X, Johannesson L, Murgovski N, Egardt B. Longevity-conscious dimensioning and power management of the hybrid energy storage system in a fuel cell hybrid electric bus. Appl Energ. 2015;**137**:913–924.

[20] Paladini V, Donateo T, de Risi A, Laforgia D. Super-capacitors fuel-cell hybrid electric vehicle optimization and control strategy development. Energ Convers Manage. 2007;**48** (11):3001–3008.

[21] Masih-Tehrani M, Ha'iri-Yazdi MR, Esfahanian V, Safaei A. Optimum sizing and optimum energy management of a hybrid energy storage system for lithium battery life improvement. J Power Sources. 2013;**244**:2–10.

[22] Nguyen A, Lauber J, Dambrine M. Optimal control based algorithms for energy management of automotive power systems with battery/supercapacitor storage devices. Energ Convers Manage. 2014;**87**:410–420.

[23] Ehsani M, Gao Y, Emadi A. Modern electric, hybrid electric, and fuel cell vehicles: Fundamentals, theory, and design. 2nd ed. New York, NY: CRC Press; 2009.

[24] Advisor version 2003. Advisor help documentation. US: National Renewable Energy Laboratory; 2013.

[25] Bataus M, Maciac A, Oprean M, Vasiliu N. Automotive clutch models for real time simulation. Proc Romanian Acad A. 2011;**12**:109–116.

[26] Malik A, Zhang Z, Agarwal RK. Extraction of battery parameters using a multi-objective genetic algorithm with a non-linear circuit model. J Power Sources. 2014;**259**:76–86.

[27] A123 datasheet of 26650m1b. US: A123 Systems Inc.; 2012.

[28] Ouyang M, Zhang W, Wang E, Yang F, Li J, Li Z, Yu P, Ye X. Performance analysis of a novel coaxial power-split hybrid powertrain using a CNG engine and supercapacitors. Appl Energ. 2015;**157**:595–606.

[29] Sorrentino M, Rizzo G, Arsie I. Analysis of a rule-based control strategy for on-board energy management of series hybrid vehicles. Control Eng Pract. 2011;**19**:1433–1441.

[30] Vinot E, Trigui R. Optimal energy management of HEVs with hybrid storage system. Energ Convers Manage. 2013;**76**:437–452.

[31] Kim N, Cha S, Peng H. Optimal control of hybrid electric vehicles based on Pontryagin's minimum principle. IEEE T Veh Technol. 2011;**19**:1279–1287.

Chapter 5

Performance Analysis of an Integrated Starter-Alternator-Booster for Hybrid Electric Vehicles

Florin-Nicolae Jurca and Mircea Ruba

Additional information is available at the end of the chapter

http://dx.doi.org/10.5772/intechopen.68861

Abstract

The chapter aims to investigate the reduction of the fuel consumption of conventional vehicles using mild-hybridization and considering the New European Driving Cycle (NEDC), using two topologies of electrical machines dedicated to integrated starter-alternator-booster (ISAB) applications: directly connected to the crankshaft (called 'normal ISAB') and indirectly through the belt system (called BSAB), respectively. The behaviour of ISAB and BSAB of a hybrid electric vehicle has been investigated with a multi-domain simulation software developed in Advanced Modelling Environment for performing Simulation (AMESim).

Keywords: electrification, hybrid electric vehicle, integrated starter-alternator-booster, electrical machine

1. Introduction

As the electrification of vehicle propulsion at low (e-bikes) and high power (buses) continues to extend, the current research efforts on this topic are focused especially on increasing the autonomy of vehicles due to the accumulation of electricity. Due to the lack of charging station and low autonomy in terms of maintaining a reduced weight of the battery, the electrical vehicle is momently limited to urban trails. In this context, the hybrid electric vehicles (HEVs) were considered initially as a transition between conventional vehicles (internal combustion engine (ICE)) and the electric ones, and now they still remain a viable solution that is gaining ground by combining the advantages of both types of vehicles [1–4].

The trend in all types of vehicles (conventional, electrical, or hybrid vehicles) for the next years is to increase the equipment with different types of electrical subsystems. These can be related

INTECH
open science | open minds

© 2017 The Author(s). Licensee InTech. This chapter is distributed under the terms of the Creative Commons Attribution License (http://creativecommons.org/licenses/by/3.0), which permits unrestricted use, distribution, and reproduction in any medium, provided the original work is properly cited.

to the safety (direction, breaking, lights, distance sensors, mirrors etc.) or to the comfort (seats, HAVC, audio, navigation display etc.). At the same time, a lot of traditional mechanically driven loads are replaced with electrical driven ones (water pumps, servo steering, ventilation fan, etc.). This demand of electrical energy, of around 10 kW [5], requires increasing generator power and a certain level of efficiency (normally situated at 40–55%) [6]. A common alternator in a car is relatively cheap and with low efficiency, but with the expected increase of power, it exceeds the capability of the Lundell generator (claw pole synchronous machine). In this context, the replacement of classical alternator with a high efficiency machine is mandatory. Besides this, the operating mode of the conventional starter (around 1 s for each start) is used only for the start of the ICE and after it becomes an extra weight in the vehicle. The easy (costs and implementation) solution of this problem is to replace both machines (starter and alternator) with a single electrical machine.

The initial concept of the integrated starter-alternator (ISA) system was developed in order to gain more space for the powertrain system and to reduce the weight of the vehicle by combining the starter with the alternator. This system ensures the start/stop of the internal combustion engine and the supply with electricity of all the auxiliary subsystems (safety or comfort).

Especially in parallel configuration of HEV, the ISA is used for starting the internal combustion and supply the electrical load. A second electrical machine is necessary for the electric propulsion. The method for the simplification of this structure involves the use of a single electric machine comprising three operating modes: starter-alternator and booster. In this case, the integrated starter-alternator-booster (ISAB) system will be able initially to start the ICE, then, when it is turned on, it will reverse to generator mode and will supply electricity to consumers and the storage system. By adopting adequate control strategies, the electrical machine is capable of moving quickly from generator to motor (booster) and back in order to help the international combustion engine for a short period of time (maximum 2–3 min), if more power is necessary (overruns, ramp, curbs, etc.) [7]. This operating mode of the machine is generically called integrated starter-alternator-booster (ISAB). Using ISAB in parallel HEV is generically called *Mild-HEV*. In this configuration, the full electric propulsion of the vehicle is not possible, but the production costs necessary for the implementation of the hybridization in conventional vehicles are lowest compared to other variants of HEV.

According to Ref. [8], where the influence of fuel consumption for a small car equipped with ISAB is investigated and considering the European standard (1999/100 EC), the fuel consumption is reduced to about 12% in total.

The increase in the number of electric components within the vehicles boosts the market for electrical motors for hybrid and electric vehicles. A Frost & Sullivan market research finds that the market earned revenues of about 55 million Euros in 2010, which are expected to reach $1.6 billion by the end of 2017 [9].

The required characteristics of the ISAB in the starter mode and alternator (generator) mode are very restrictive for a conventional electrical machine [10]:

http://dx.doi.org/10.5772/intechopen.68861

- High value of electromagnetic torque for starting ICE (120–300 Nm).
- High efficiency at wide speed range (600–6000 rpm).
- Reliability at high vibrations and over 250,000 start/stop cycles (during 10 years).
- Operation at temperatures between −30 and 130°C.
- Easy maintenance and low cost and so on.

Usually, the ISAB can be connected with a gasoline or diesel engine, either directly through the crankshaft (see **Figure 1**) or indirectly through the belt system (see **Figure 2**); based on that, the systems are called belt-driven starter-alternator-booster (BSAB) and conventional ISAB, respectively.

The size of the electrical machine is very important for BSAB application in order to keep the overall size low (the same dimensions like the ones of a conventional alternator) but for a given maximum torque, the systems usually have a recommended gear ratio 3:1 to the ICE crankshaft, according to Ref. [8]. The BSAB runs with a speed three times higher than the ICE. For the ISAB, the speed range is usually synchronized with combustion engine.

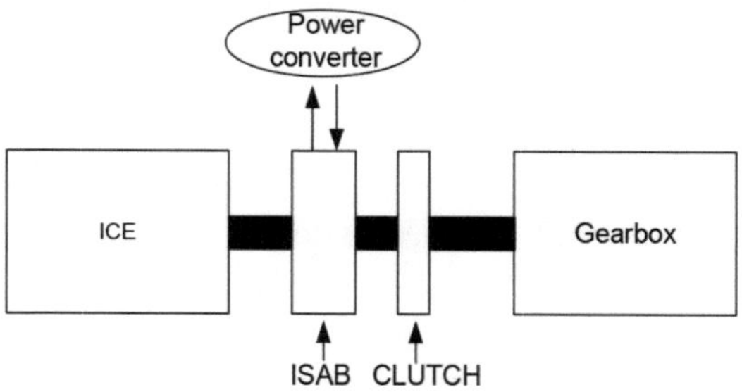

Figure 1. The ISAB system.

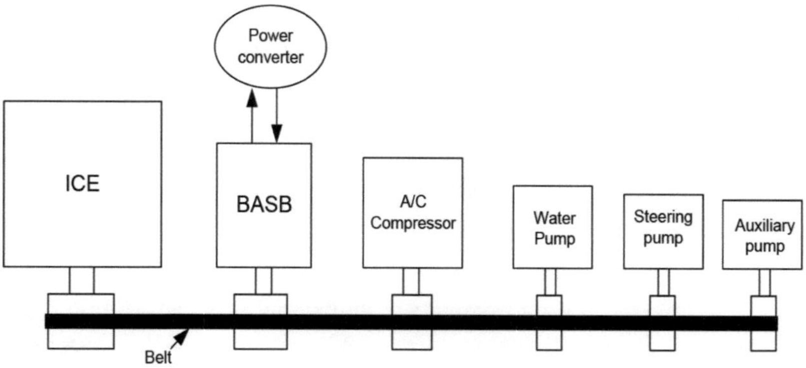

Figure 2. The BSAB system.

2. Electrical drive used for ISAB applications

2.1. Electrical machines

In the last decade, the development of power electronics (inverter/convertor) made the alternative current (AC) machines the best solution for ISAB applications, especially due to their high power density. These are synchronous reluctance machine (SynRM), induction machine and permanent magnet synchronous machines (PMSM) in both supplying variants: with sinusoidal and trapezoidal current.

The detailed investigation of SRM and induction machine is presented in Refs. [11, 12]. In these studies, the complicated electronics needed for SRM and the difficult control of the induction machine (influence of slip in performance of the machine) are highlighted. In this context, the SynRM and PMSM are the best candidates for ISAB applications.

The electrical machines used for conventional ISAB applications are exposed at high temperatures generated by ICE. This makes impossible the use of the PMSM in high efficiency and low-cost conditions (only with a special method for cooling or using expensive SmCo magnet). Therefore, the SynRM without permanent magnets is the best solution for the direct connection to the crankshaft of ICE (ISAB) and PMSM machine for BSAB.

2.1.1. PMSM machine for BSAB applications

The main advantage of the PMSM compared with other types of electrical machine is their high efficiency due to the absence of the field coil losses. The stator is constructed from three-phase windings and steel sheets (the same as the induction machine), but due to the absence of iron losses, the rotor is built from massive steel and permanent magnets. The position of the permanent magnets can be categorized as surface-mounted type and interior type. This position can have a significant effect on the mechanical and electrical characteristics, especially on the synchronous inductance [13]. Because the permeability value of rare earth magnet (such as NdFeB) is very close to that of the air, the air gap of the machine with mounted surface PM effectively becomes larger in this case. This makes the machine *d*-axis inductance value very low, with a significant effect on the ability of overloading the machine and operation at flux weakening. Because the maximum torque is inverse proportional with the *d*-inductance, this becomes very large. But the low value of *d*-inductance reduces the possibility to operate at flux weakening. This is caused by the need to use a high value of the demagnetization component of the stator current in order to decrease the flux value in the air gap. Therefore, the remained current on the *q* axis will be insufficient to produce torque.

In the case of the interior magnets, it is possible to obtain a sinusoidal distribution of the air-gap flux by using simple rectangular magnets. A sinusoidal flux distribution reduces considerably the cogging torque, in particular in the case of the machine with a large number of pole pairs and a small number of slots per pole and phase [14]. For these structures, it is also possible to increase the flux density in the air gap beyond the value of the remnant flux density of the magnets by using the flux concentrators. Because in this case the *d*-inductance is usually higher than with that of the surface magnets topologies, the overload capacity of the machine will be reduced and the performance in flux weakening conditions will be higher.

The PMSM with outer rotor (PMSMOR) (see **Figure 3**) is one of the special topologies of PMSM, with some advantages for BSAB applications:

- Belt mounted directly on the outer rotor, without using pulley.
- Easy mounting of permanent magnets, the centrifugal forces do not influence their mechanical stability.
- High torque capabilities.
- Convenience of cooling, etc.

The development cycle of PMSM (inner or outer rotor) topologies includes analytical procedure, magnetic field analysis and optimization procedure connected to previous design steps. The analytical procedure is presented in detail in Refs. [13, 14] and includes the following topics: analysis of the specifications, selection of the topology, the active and passive materials, sizing the machine, choice of the manufacturing technologies and information about preliminary cost evaluation. In the dimensioning procedure, classical formulas or dedicated software platforms like SPEED software can be used.

The electromagnetic flux analysis is realized with dedicated programs (like Flux 2D/3D, Jmag 2D/3D, Maxwell 2D/3D, ANSYS, Opera, open-source programs, etc.) based on the finite element method (FEM). The FEM is a widely used method for obtaining a numerical approximate solution for a given mathematical model of the machine. The obtained results are related to the voltage/current waveform, map of flux density, electromagnetic torque, losses (iron and Joule), and the efficiency value or map of it.

The optimization of electric machine is a multivariable, nonlinear problem with constraints. In order to treat problems with constraints, it is necessary to transform them into unconstrained ones. This can be done, for instance, by embedding the constraints in the objective function.

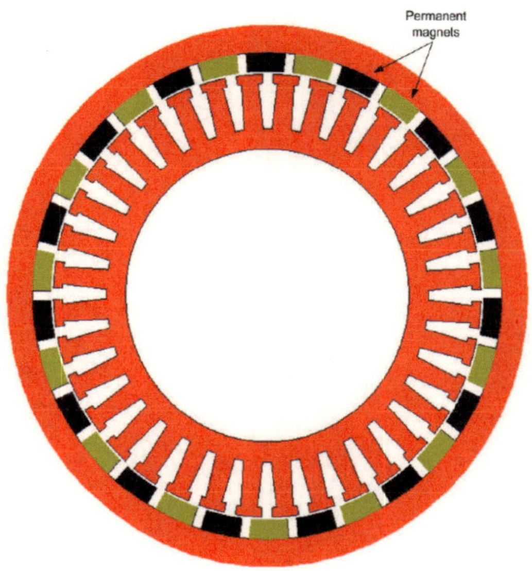

Figure 3. PMSMOR type.

The most used optimization algorithms in design of all types of electrical machines are as follows: genetics algorithms (GA), differential evolution algorithm (DEA), estimation of distribution algorithms (EDAs), particle swarm optimization (PSO) and multi-objective genetic algorithms (MOGA, Pareto, etc.) [15].

A comprehensive evaluation of optimization algorithms was performed in Refs. [16–18]. The authors of these studies state that any such classification of different optimization algorithms is not truly appropriate since the performance is an objective closely related to the specifics of each application. Nevertheless, in the optimization of the electrical machine, the authors mostly agree that DEA achieves the best fitness values, i.e. the minimum objective function value, usually with a smaller number of evaluation steps.

Considering the important step in the development of cycle of PMSM presented above, a general design procedure of PMSMOR for BSAB applications is proposed and presented in **Figure 4**.

2.1.2. SynRM machine for ISAB applications

Variable reluctance synchronous machines have received little attention in various comparative studies approaching the selection of the most appropriate electric-propulsion system for either HEV or EV. Malan [19, 20] showed that the SynRM drive has major advantages in electrical

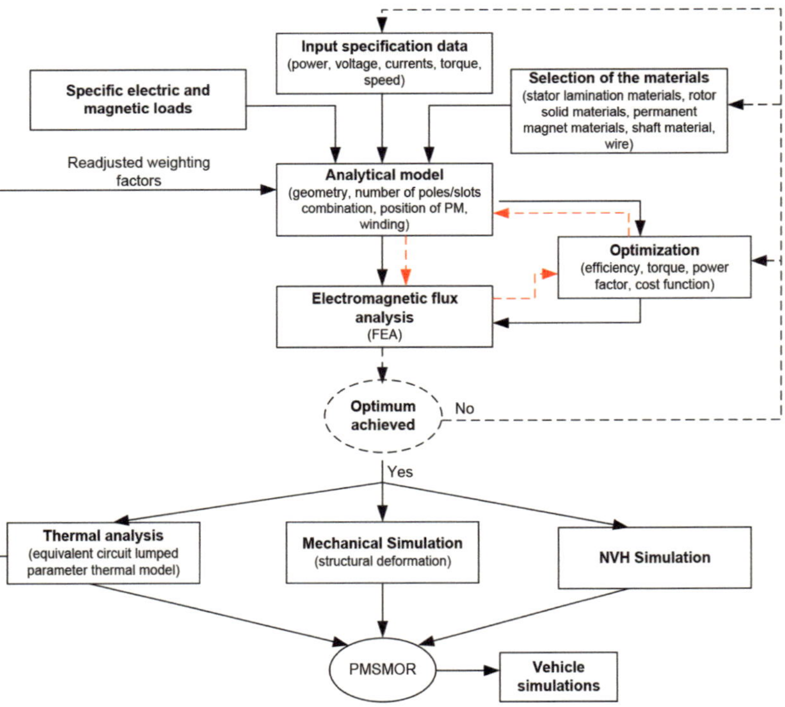

Figure 4. General design procedure of PMSMOR.

propulsion. SynRM's performance strongly depends on the saliency ratio, but increasing the saliency complicates the rotor construction and drastically increases the motor cost. Interesting results concerning the influence of the saliency ratio on the SynRM steady-state performances, mainly on power factor and efficiency, are given in Ref. [21], while the effect of rotor dimensions on *d*- and *q*-axis inductances in the case of a SynRM with flux barrier rotor is discussed in Ref. [22]. Thus, the number of rotor flux barrier for the SynRM recommended in the literature is four. Above this value, the technology of the rotor is too complicated, while for a value lower than 4, the value of the torque ripple is too high. Regarding the rotor construction, there are three main different types, given in Ref. [23], presented in **Figure 5**.

- With salient rotor poles (see **Figure 5a**): require low technological effort and are obtained by removing the iron material from each rotor pole in the transversal region.
- With axially laminated rotor (ALA) (see **Figure 5b**): the rotor core is made of axial steel sheets that are insulated from each other using electrically and magnetically insulation (passive material).
- Transversally laminated anisotropic rotor (TLA) (see **Figure 5c**): the so-called ribs are obtained by punching and then the various rotor segments are connected to each other by these ribs.

The SynRM has a larger torque density compared with that of IM. This comes from the absence of rotor cage and related losses. A different dynamic behaviour is expected from SynRM due to the specific relationships between currents and fluxes. Because SynRM does not have a traditional cage (especially used for starting), it is necessary to use the modern inverter technology. Therefore, most of the literature on SynRM drives has concentrated mainly on the design and control of the machine with the goal of improving control, efficiency and torque production, drive flexibility and cost [24].

The main drawback of SynRM is related to structural behaviour at high speeds (over 10,000 rpm) because the specific geometry of the rotor involves thin layers of steel and large cut-off areas.

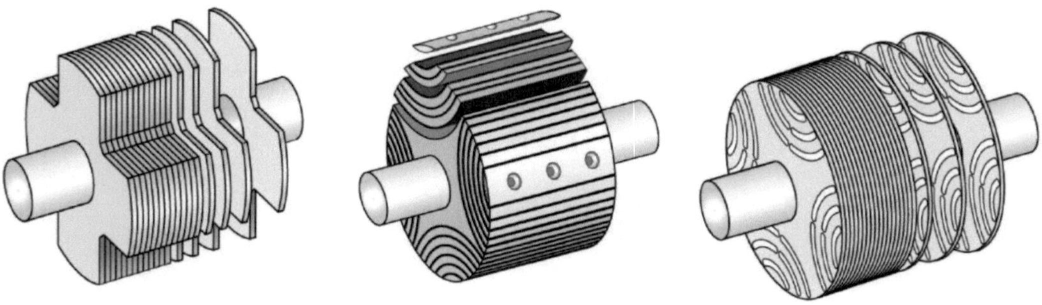

(a) (b) (c)

Figure 5. Rotor topologies for a SynRM: (a) simple salient pole, (b) axially laminated anisotropy rotor, and (c) transversally laminated anisotropic rotor. *Source*: [23].

Based on advantages and disadvantages of the SynRM and the specific applications of ISAB (rated speed up to 10,000 rpm), SynRM is one of the most suitable candidates for direct connection to crankshaft. The major advantages are high torque, thermal behaviour (absence of permanent magnets and low average value of iron losses), high value of efficiency at entire drive cycle of functioning, vibro-acoustic behaviour (low noise), etc.

In the development cycle of SynRM presented below, the most important step is related to the rotor geometry and the structural behaviour (see **Figure 6**).

2.2. Power electronics

The electrical equipment installed on the vehicles operates at a nominal voltage of 14 V. In the early 1990s, a new standard (PowerNET) for automotive electrical systems has been proposed by a consortium of automotive manufacturers (Daimler-Benz and General Motors). Following a proposal by the PowerNET, the voltage level increases for the electrical installation to 42 V [25]. The goal was to reduce the section of the conductors and gain the possibility to increase the total

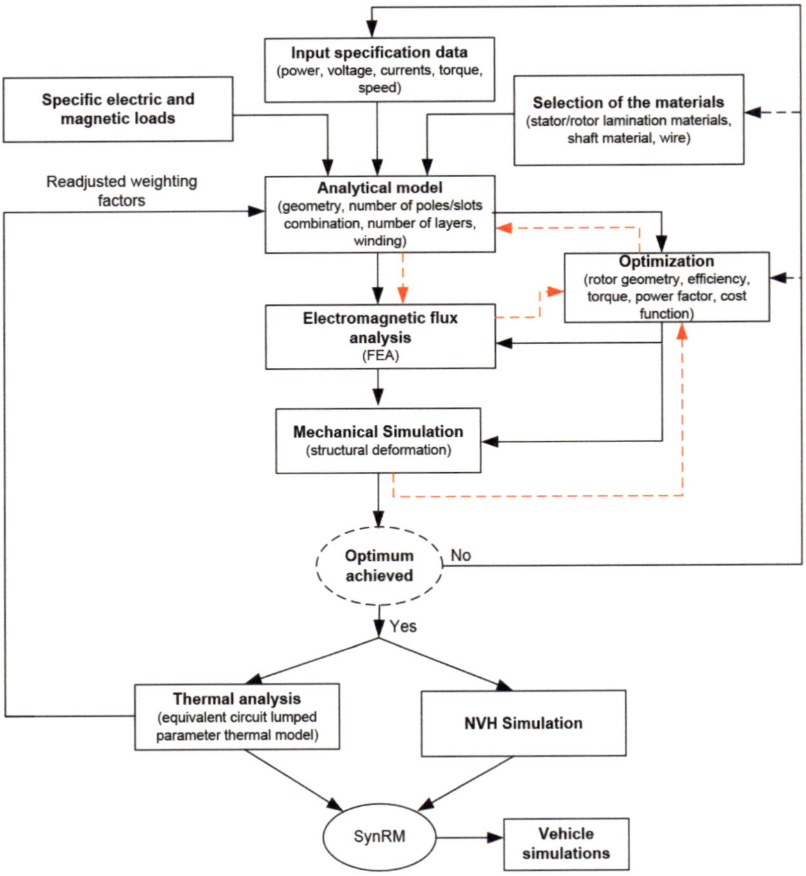

Figure 6. General design procedure of SynRM.

power installed in the new generation of vehicle. The standard did not become very popular because of its high implementation costs, which would require the redesign of all electrical and electronical subsystems [26]. Instead, most producers were oriented on systems with two voltage levels: high voltage for propulsion and low voltage for auxiliary and electronic subsystems.

A starter-alternator system involves the use of a static frequency converter for the driving of the electrical machine. The convertor will operate in both the inverter and rectifier regimes. In the rectifier operating mode, it is indicated to adopt a control strategy of the converter with the purpose of reducing losses and the harmonic content of the output currents of the machine. The techniques for the control of the converter for these two modes are the same, only the current reverses its sense depending on the operating mode.

The input voltage of the static frequency converter is a DC voltage, the value of which must be kept constant in order to function optimally. The regulation of the input voltage of the converter can be done by using a bidirectional DC/DC converter with a closed loop control. An alternative to the use a DC/DC stage converter and another DC/AC converter is to use a Z-Source Converter [27]. The Z-Source Converter is more capable compared with the classical converter to operate both as a boost and buck converter due to the input impedances that give it particular operating properties.

2.2.1. *Power electronics of SynRM and PMSM*

For the control of PMSM machine, the current of the *q* axis is maintained maximum in order to produce high value of the torque and zero for *d* axis current, respectively. Instead, for SynRM, the control strategies mean to keep the equal value of the *q* axis current with the *d* axis. In the case of PMSM with interior magnets, this control strategy does not provide maximum torque due to the additional reluctant torque [28] component that appears in expression:

$$T = \frac{3}{2} \cdot p \cdot \left[\Psi_{PM} \cdot i_q - (L_q - L_d) \cdot i_d \cdot i_q\right] \tag{1}$$

where T is the electromagnetic torque, p is a pair pole number, Ψ_{PM} is the permanent magnet flux, i_q is q axis current, i_d is d axis current, and L_q, L_d are q axis and d axis inductances.

The reluctant component of the torque has a maximum value for $i_d \neq 0$ and the stator current equal with $\pi/4$.

Usually, the implemented control method for the PMSM and SynRM for automotive application is an indirect method, which is based on measuring the stator currents and calculating the rotor flux phasor magnitude and position using these currents and the rotor position. Thus, the flux transducer or flux estimators that are usually used in the vector control method with direct measurement of flux are eliminated. This method has a disadvantage due to the fact that the accurate determination of rotor flux phasor position requires a precise measurement of rotor position. Thus, the practical implementation using speed measurement for obtaining the integration of the rotor angle is not recommended. Hence, an incremental encoder position or a resolver, which has a higher cost while providing the precision required of a vector control with a good dynamic response in applications is used. In addition to this vector control method that uses position sensors for determining the rotor angle control, other methods where these transducers are eliminated (sensorless) exist. In these cases, the rotor position is estimated by using complex algorithms, using as input the measurement values of voltages and currents [29].

The general diagram control presented in **Figure 7**, usually used for PMSM and SynRM, can be divided into power and control components. The power circuit consists of the electrical machine (PMSM/SynRM), DC/DC converter, inverter, while the control loop consists of speed transducer, current transducers, PWM signal generation block, transformation of coordinate systems blocks and computing block of references current.

The control strategies considered for the SynRM are:

- Maximum torque control per ampere control (MPTAC)

The model of control is based on imposing the same currents for the *d* and *q* axes of the machine as current references for the vector control of the machine. These currents are calculated from the torque equation like

$$i_d = i_q = \sqrt{\frac{T_{\text{ref}}}{p \cdot (L_d - L_q)}} \tag{2}$$

- Maximum rate of change of torque control (MRCTC)

The control strategy is implemented in order to obtain fast machine response at sudden torque steps of the load. The idea here is to compute the *d* and *q* current component functions of some machine parameters and to use these components as input of the vector control scheme. The detailed control strategy is presented in Ref. [30].

- Maximum power factor control (MPFC)

The aim of this method is to maximize the power factor of the machine. For this, the *d* and *q* current components are computed as follows:

$$I_{q\text{imposed}} = T_{\text{ref}} \cdot \frac{1}{p} \cdot \frac{1}{L_d - L_q} \cdot \frac{1}{I_{d\,(\text{from the machine})}} \tag{3}$$

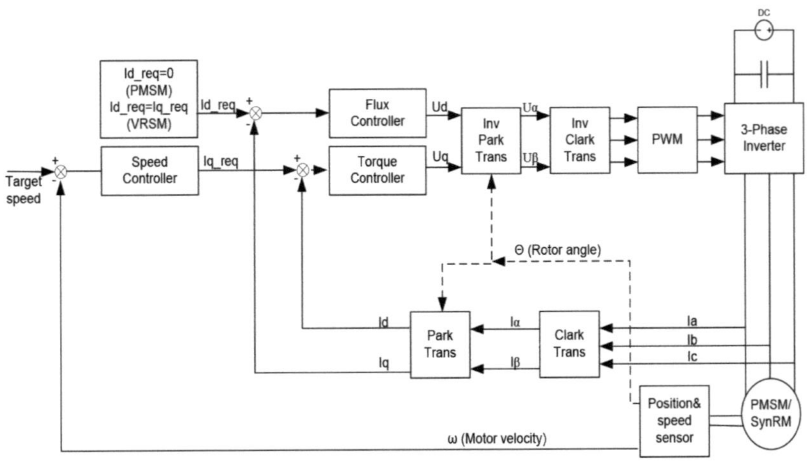

Figure 7. Block diagram of vector control system for PMSM and SynRM.

$$I_{dimposed} = I_{qimposed} \cdot \frac{\cos(\theta)}{\sin(\theta)} \tag{4}$$

For the ideal machine, θ is in the range 0–90° and gives best performance at 45°. For a real machine, it has to be computed by varying its value and observing the performance of the machine for each angle step.

3. Simulation of a hybrid electric vehicle with the ISAB system

In order to study the electrical machines in ISAB applications, the electric drive model can be introduced and simulated in the Advanced Modelling Environment for performing Simulation (AMESim). AMESim is a multi-domain simulation software for the modelling and analysis of one-dimensional (1D) systems. In this program, each component or physical phenomenon is described by differential equations, type formulation in which the major variable is the time [31]. This approach is different from the partial derivate equations formulation, which formalizes the notion of the distribution of system properties in space. The representation of a dynamic system starting from the notion of "multiport" consists of highlighting the energy exchanges between a component and its environment through the connecting ports. The connection of two or more components through the port allows port exchange power (electrical, mechanical, etc.) according to the adopted sign convention.

For automotive applications, the program comprises discrete components of the ICE, gearbox, control system, electric loads, electrical machine and power inverter, connected together to form a global model of a hybrid electric vehicle.

The geometrical and electrical parameters of electrical machine considered for ISAB (SynRM) and BSAB (PMSMOR) application are presented in **Table 1**. The configuration of PMSMOR is a three-phase machine with 36 slots and 15 poles, and the SynRM topology is a three-phase machine with 27 slots and 4 poles. The dimension of PMSMOR has been imposed according to Ref. [32] (data chosen for belt brushless alternator).

The simulation of the BSAB and ISAB is carried out on a New European Driving Cycle (NEDC). A driving cycle is a series of points defining a speed profile that the studied vehicle must follow [33]. The defined speed profile simulates most common operating modes of an automobile (frequent acceleration and deceleration, load variations and speed variations) and corresponds to both urban and extra-urban environments. The parameters and the profile of NEDC are presented in **Table 2** and **Figure 8**, respectively.

The model takes into account the most complex thermodynamic phenomena occurring in a heat engine. In the initial implementation, the starter and alternator were a DC permanent magnets machine and the Lundell generator, respectively. The model has been replaced by the studied model and is shown earlier (PMSMOR/SynR). The motors are powered from a battery through the converter DC/DC that will operate in this case as a boost converter.

The evaluation of the performance of PMSMOR and SynRM was started from a demonstration model in AMESim of a compact car category (see **Figures 9** and **10**) with a compression-ignition

Parameter	PMSMOR	SynRM
Output power (W)	6500	10,000
Rated speed (rpm)	400	800
Phase voltage (V)	72	100
Number of phases (–)	3	3
Number of pole pairs (–)	15	2
Number of slots	39	30
Stator outer diameter (mm)	176	260
Stack length (mm)	150	150
Rotor outer diameter (mm)	210	210
Tooth width (mm)	5	6
Permanent magnet height (mm)	8	–
Residual flux density—NdFeB-N48 (T)	1.4	–
Coercive force—NdFeB-N48 (kA/m)	796	–
Stator and rotor (only for SynRM) lamination type	M270-35A	
Rated current (A)	72	72
Current density (A/mm^2)	7	6
Iron losses (W)	212	623
Torque (N m)	150	150
Power factor	0.9	0.85
Efficiency (%)	90	87
Saliency ratio	–	4.1

Table 1. Geometrical and electrical parameters of PMSMOR and SynRM.

Time (s)	Distance (km)	Max speed (km/h)	Average speed (km/h)	Max acceleration (m/s^2)	Max deceleration (m/s^2)	Idle functioning (s)	Stops
1184	10.93	120	33.21	1.06	-1.39	298	12

Table 2. NEDC parameters.

combustion engine. The imposed weight of the vehicle was 1200 kg (usually between 1134 and 1360 kg, according to Ref. [34]) without any extra weight or passengers.

In order to have comparative results regarding the fuel consumption, in the first simulation, the conventional vehicle functioning during the NEDC cycle was tested. In the next simulations,

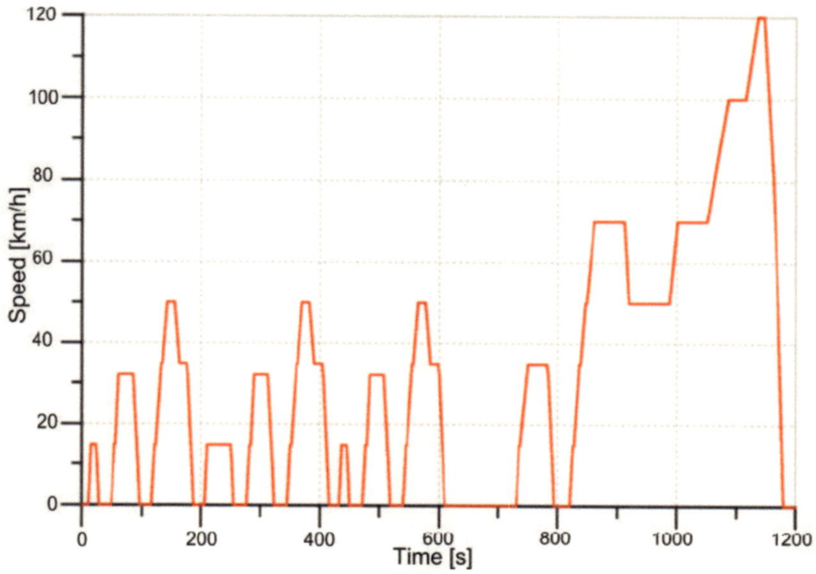

Figure 8. NEDC profile.

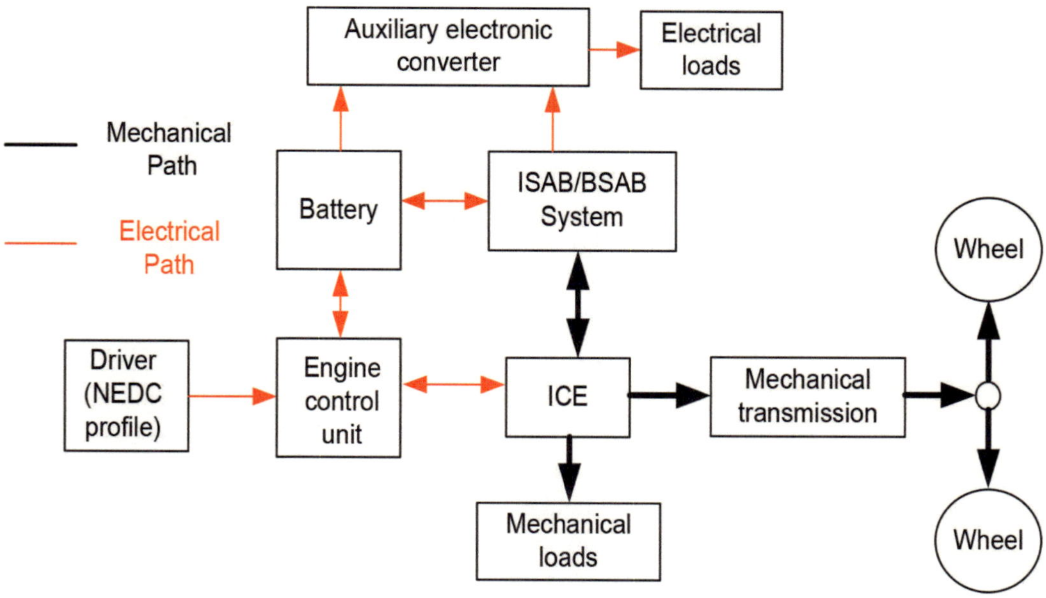

Figure 9. Structural diagram of the vehicle.

the ISAB regime with considered electrical machines was established. The behaviour of the starter and alternator in the vehicle model was supposed to be the same as in a conventional car. For the booster regime, a set-up to help ICE for 15 min/h was added and this works in the booster regime only when the battery was fully charged (up to 95%). The time limit for each

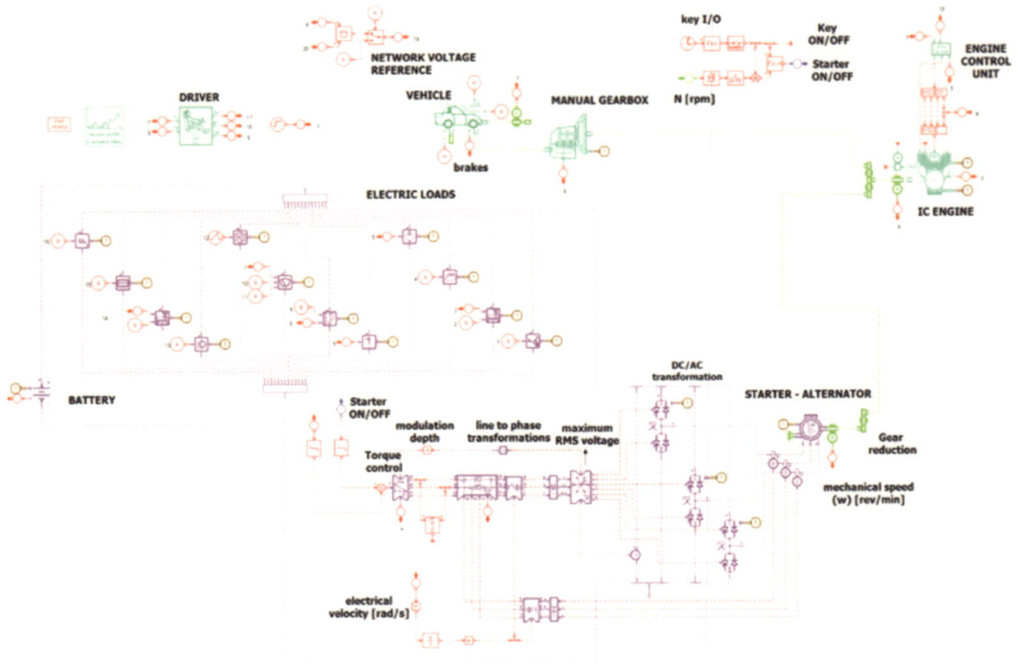

Figure 10. Vehicle model [35].

booster regime was set at 2 min in order to avoid the complete discharging of the battery (but no less than 20%). The parameters of the ICE considered are presented in **Table 3**.

The control of the electrical machines in the starter operating mode involves a maximum torque value (150 N m) until the ICE reaches 350 rpm. Torque command is provided by a bi-positional regulator with hysteresis. It is active when the command of ICE is active and its speed is less than 200 rpm, and it is off when the speed exceeds 400 rpm. When the ICE speed exceeds 300 rpm, the process of fuel injection into the cylinders starts and the ICE accelerates to idle speed. By applying the necessary torque to start the ICE, this is accelerated rapidly at the speed of 400 rpm in about 0.35 s.

When ICE exceeds the speed of 400 rpm, the bi-positional controller becomes inactive and the combustion engine continues to spin out due to its inertia. If the pistons do not reach the maximum compression point, they will not be able to inject fuel to start the combustion process; consequently, the speed drops below 200 rpm and now the controller output is active. Therefore, the starter is controlled again and the combustion engine is brought to a speed of 400

Type	**Turbo diesel**	**Compression ratio**	**2:1**
Number of cylinders	4	Maximum torque	365 N m at 2000 rpm
Cubic capacity	1994 cm^3	Maximum power	120 kW at 3750 rpm

Table 3. ICE parameters.

rpm. At the start of the combustion engine process, ICE is accelerated to the idle speed, where it is maintained by the electronic control unit. The entire process of starting the engine (in normal condition), from the beginning until stable operation at idle speed, lasts 0.8 s. In the winter, this process may take 1.5 s. The speed profile of starting the ICE is presented in **Figure 11**.

For the alternator mode, the nominal value of the electrical loads is considered in the model. Some electrical loads are intermittently connected (fan, electrical window, heating systems, etc.). Other loads are dependent on ICE speed (fuel pump and injectors) or the speed of the vehicle.

When the entire driving cycle is considered, the fuel consumption in the vehicle is reduced to 878.63 ml for the BSAB system and 941 ml for the ISAB system. These values represent a fuel economy of around 16% for BSAB and 17.3% for ISAB of total consumption compared with a classical vehicle with a dedicated alternator and starter (without booster option) system. The difference in fuel consumption is due to the value of nominal power of electrical machines (see **Table 1**). But, the performances of SynRM are limited by the battery (which uses 75 Ah) capacities. If it uses a stronger battery, the total fuel economy can be increased with 2 or 4% (especially due to the booster mode).

In the mechanical evaluation of electrical machines for automotive applications, the variation of electromagnetic torque is one of the most important parameters, because this variation (torque ripple) can become a source of noise and vibration in vehicles. Thus, for a better visualization of the torque profile of PMSMOR and SynRM, a new scenario for all three regimes was considered. For better comparative results (variation of axis torque) between IASB and BSAB, the BSAB system is taken into account through directly coupling (using ratio 1:1 between ICE and BSAB speed) at ICE. The starter and generator regime has been set for 1.5 and 20 s, respectively. The variation of the axis torque in the generator mode has been obtained by intermittent connection of the electrical loads (lights, HVAC, media, etc.). For the booster mode, the speed of the vehicle is increased from 70 to 120 km/h in 17 s, necessary to overtake other vehicles. In this case, the battery is considered fully charged.

Figures 12 and **13** show the variation of the axis torque versus time for all operating modes, respectively. Due to the proper windings-slot combination, the torque ripple values are below

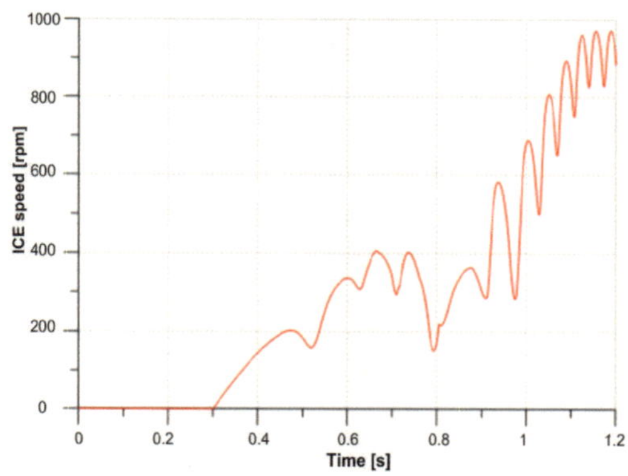

Figure 11. Speed profile at starting ICE.

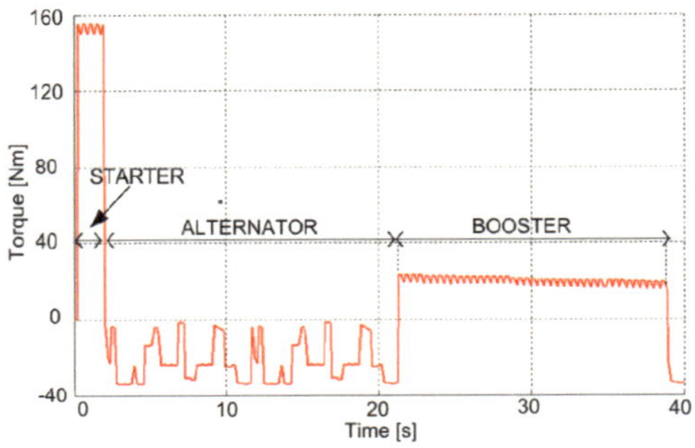

Figure 12. BSAB torque profile.

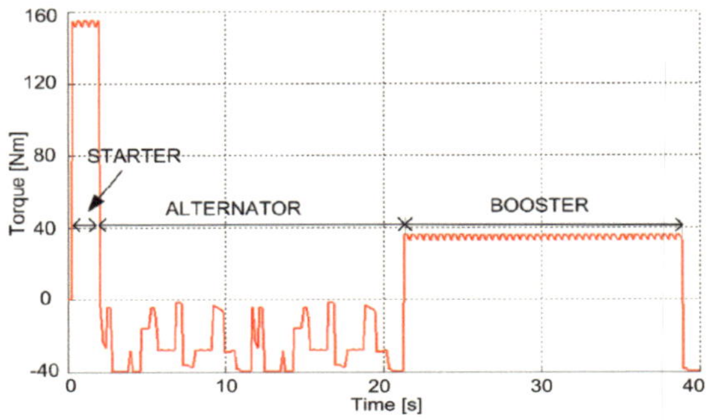

Figure 13. ISAB torque profile.

10%. In fact, the ratio of the torque ripple is 7.1% for PMSMOR and 6.2% for SynRM. In the booster mode, the rated torque value of the 10 kW SynRM machine used for ISAB is obviously bigger than that of the 6.5 kW PMSMOR.

4. Conclusions

The chapter presents the main steps to be followed in the development of a specific electric drive system dedicated to automotive domain, such as integrated starter-alternator-booster applications. Replacing the starter and alternator in a conventional vehicle with extended possibility to work in booster mode represents the first step of vehicles' hybridization, called *Mild-HEV*. In this way, two variants in mounting the ISAB had been identified in the literature: one, directly driven generic called ISAB and another belt-driven called BSAB. In this chapter, the approach contains the major elements that need to be discussed for two type of electrical machine (PMSMOR and SynRM) suitable for BSAB and ISAB, respectively. The general design procedure is presented for these two electrical machines by taking into account the typical constraints of the applications and the behaviour of the machine (thermal, structural, and noise, vibrations and harshness particularities). Also, the control aspects of both electrical machines are presented.

In order to demonstrate the capability of this vehicle hybridization method, two electrical machines have been designed and the equations model was developed and implemented in the general 1D model of conventional vehicle performed in AMESim software. The influence of fuel consumption on the entire drive cycling (NEDC) was investigated. Based on the obtained results, the ISAB system gives a greater value of the reduction of fuel consumption, but the coupling of the electrical machine directly to the crankshaft involves complicated manufacturing techniques (higher cost) compared with the BSAB system procedure.

Acknowledgements

This work was supported by a grant of Strengthening the Research potential of CAREESD in the field of Electromechanical Systems and Power Electronics for Sustainable Applications, ESPESA, 692224/2015 H2020 TWINNing-2015 and ALNEMAD (PCCA 181/2012).

Author details

Florin-Nicolae Jurca* and Mircea Ruba

*Address all correspondence to: florin.jurca@emd.utcluj.ro

Department of Electrical Machines and Drives, Technical University of Cluj-Napoca, Cluj-Napoca, Romania

References

[1] Andreescu G-D, Coman CE. Integrated starter-alternator control system for automotive. In: Proceedings of the International Symposium on Computational Intelligence and Informatics (CINTI '13), 19-21 Nov. 2013, Budapest, Hungary, p. 339-343, Publisher: IEEE.

[2] Mirahki H, Moallem M, Rahimi SA. Design optimization of IPMSM for 42 V integrated starter alternator using lumped parameter model and genetic algorithms. IEEE Transactions on Magnetics. 2014, Vol: 5, NO: 3, pp. 1-6, DOI: 10.1109/TMAG.2013.2285358

[3] Rizoug N, Barbedette B, Sadoun R, Feld G, Starter-alternator propel the vehicle through a hybrid supply: Battery and supercapacitors. In: Proceedings of the Twenty-Seventh Annual IEEE Applied Power Electgronics Conference and Exposition (APEC'12); 5-8 February 2012; Orlando, Florida, USA. pp. 2583-2589, Publisher: IEEE.

[4] Mirahki H, Moallem M. Design improvement of Interior Permanent Magnet synchronous machine for Integrated Starter Alternator application. In: Proceedings of the International Electric Machines & Drives Conference (IEMDC'13); 12-15 May; Chicago, Illinois, USA. pp. 382-385, Publisher: IEEE.

[5] Parsa L, Goodarzi A, Toliyat HA. Five-phase interior permanent magnet motor for hybrid electric vehicle application. In: Proceedings of the IEEE Conference of Vehicle Power and Propulsion (VPPC' 05); 7-9 September 2005; Chicago, Illinois, USA. pp. 631-637, Publisher: IEEE.

[6] Ivankovic R, Cros J, Kakhki MT, Martins CA, Viarouge P. Power electronic solutions to improve the performance of Lundell automotive alternators. In: Carmo J, editors. New Advances in Vehicular Technology and Automotive Engineering. Rijeka, InTech; 2012. DOI:10.5772/48459

[7] Jurca FN, Ruba M, Martis C. Analysis of permanent magnet synchronous machine for integrated starter-alternator-booster applications. In: Proceedings of the 2015 International Conference on Electrical Drives and Power Electronics (EDPE'15); 21-23 September 2015; High Tatras, Slovakia. pp. 272-276, Publisher: IEEE.

[8] Hagstedt D. Comparison of different electrical machines for belt driven alternator starters [thesis]. Lund: Department of Measurement Technology and Industrial Electrical Engineering, Lund University; 2013

[9] Frost & Sullivan, Strategic Analysis of Electric Motor Technologies for Electric and Hybrid Vehicles in Europe. http://www.automotive.frost.com [Accessed: 6-January 2017]

[10] Bae BH, Sul S.K, Practical design criteria of interior permanent magnet synchronous motor for 42V integrated starter-generator, In: Proceedings of the IEEE International Electric Machines and Drives Conference (IEMDC' 03); 1-4 June 2003; Madison, Wisconsin, USA. pp. 656-662

[11] Schofield N, Long S, Generator operation of a switched reluctance Starter/Generator at extended speeds. IEEE Transactions on Vehicular Technology. 2009;**58**:48-56. DOI: 10.1109/TVT.2008.924981

[12] Rehman H. An integrated starter-alternator and low-cost high performance drive for vehicular applications. IEEE Transactions on Vehicular Technology. 2008;**57**:1454-1465. DOI: 10.1109/TVT.2007.909255

[13] Krishnan R. Permanent Magnet Synchronous and Brushless DC Motor Drives. Boca Raton, FL: CRC Press/Taylor & Francis; 2010. p. 611

[14] Gieras JF, Permanent Magnet Motor Technology: Design and Applications. 3rd ed. Boca Raton, FL: CRC Press/Taylor & Francis; 2011. p. 608

[15] Lei G, Zhu J, Guo Y. Multidisciplinary Design Optimization Methods for Electrical Machines and Drive System. Berlin/Heidelberg:/Springer-Verlag; 2016. p. 251

[16] Mutluer M, Bilgin O. Comparison of stochastic optimization methods for design optimization of permanent magnet synchronous motor. Neural Computing and Applications. 2012;**21**(8):2049-2056

[17] Deb A, Gupta B, Roy J. Performance comparison of differential evolution, genetic algorithm and particle swarm optimization in impedance matching of aperture coupled microstrip antennas. In: 11th Mediterranean Microwave Symposium (MMS); September 2011; Tunisia. pp. 17-20, Publisher: IEEE.

[18] Stipetic S, Miebach W, Zarko D. Optimization in design of electric machines: Methodology and workflow; In: Proceedings of the ACEMP-OPTIM-ELECTREMOTION join conference; 2-4 September 2015; Side, Turkey. pp. 1-8, Publisher: IEEE.

[19] Malan J, Kamper MJ. Performance of a hybrid electric vehicle using reluctance. IEEE Transaction on Industry Applications. 2001;**37**:1319-1324. DOI: 10.1109/28.952507

[20] Jurca FN, Ruba M, Martis C. Design and control of synchronous reluctances motors for electric traction vehicle. In: Proceedings of the 2016 International Symposium on Power Electronics, Electrical Drives, Automation and Motion (SPEDAM '16); 22-24 June 2016; Anacapri, Italy. pp. 1144-1148, Publisher: IEEE.

[21] Wu H, Lin Q, You L. An investigation of the synchronous motor. In: Proceedings of the Fifth International Conference on Electrical Machines and Systems (ICEMs' 01); 18-20 August 2001; Shenyang, China. pp. 148-151, Publisher: IEEE.

[22] Wang K, Zhu ZQ, Ombach G, Koch M, Zhang S, Xu J, Optimal slot/pole and flux-barrier layer number combinations for synchronous reluctance machines. In: Proceedings of the Eight International Conference and Exhibition on Ecological for Synchronous Reluctance Machines (EVER' 13); 27-30 March 2013; Monte Carlo. pp. 1-8, Publisher: IEEE.

[23] Fukami T, Momiyama M, Shima K, Hanaoka R, Takata S. Steady-state analysis of a dual-winding reluctance generator with a multiple-barrier rotor. IEEE Transactions on Energy Conversion. 2008;**23**:492-498. DOI: 0.1109/TEC.2008.918656

[24] Bianchi N, Blognani S, Carraro E, Castiello M, Fornasiero E, Electric vehicle traction based on synchronous reluctance motors. IEEE Transaction on Industry Applications. 2016;**52**:4762-4769. DOI: 10.1109/TIA.2016.2599850

[25] Ehsani M, Gao Y, Gay S. Characterization of electric motor drive for traction applications. In: Proceedings of the IEEE 29th Annual Conference of the Industrial Electronics Society (IECON'03); Roanoke, USA. pp. 891-896

[26] Hangiu R.P, Martis C. A Review of Automotive Integrated Starter Alternators, The Scientific Bulletin of Electrical Engineering Faculty, 2012, p.1-6

[27] Yamanaka M, Koizumi H. A bi-directional Z-source inverter for electric vehicles. In: Proceedings of the International Conference on Power Electronics and Drive Systems (PEDS'09); 2-5 November 2009; Taipei, Taiwan. pp. 574-578, Publisher: IEEE.

[28] Cai W. Comparison and review of electric machines for integrated starter alternator applications. In: Proceedings of the 39th IAS Annual Meeting; 3-7 October 2004; Seattle, USA. pp. 386-393, Publisher: IEEE.

[29] Novotny DW, Lipo TA. Vector Control and Dynamics of AC Drives. Oxford: Clarendon Press; 1999. p. 464

[30] Ruba M, Jurca FN, Martis C. Analysis of Synchronous Reluctance Machine for Light Electric Vehicle Applications, In: Proceedings of the 2016 International Symposium on Power Electronics, Electrical Drives, Automation and Motion (SPEDAM '16), 22-24 June 2016, Anacapri, Italy, pp. 1138-1143

[31] AMESim® User's Guides [internet].https://www.plm.automation.siemens.com/en/products/lms/imagine-lab/amesim. Accessed: 2017-02-27.

[32] Electrical Specifications & Selection Guide Starter and Alternators-Delco Remy. 2008 [internet]. http://www.dieselusa.com/productinfo/Delco%20Electrical%20Specs%20and%20Seletion%20Guide.pdf. Accessed: 2016-08-12.

[33] Zhang X, Chris M. Vehicle Power Management. Berlin/Heidelberg: Springer-Verlag; 2011. p. 346

[34] Code of Federal Regulations, Title 40, Protection of Environment, Parts 425 to 699. US: Office of the Federal Register; 2010

[35] Jurca FN, Martis C, Hangiu RP. Design and performance analysis of an integrated starter-alternator for hybrid electric vehicles. In: Proceedings of the Interdisciplinary research in engineering: Steps towards breakthrough innovation for sustainable development. Advanced Engineering Forum. 2013;**8-9**:453-460

Chapter 6

Design, Optimization and Modelling of High Power Density Direct-Drive Wheel Motor for Light Hybrid Electric Vehicles

Ioannis D. Chasiotis and Yannis L. Karnavas

Additional information is available at the end of the chapter

http://dx.doi.org/10.5772/intechopen.68455

Abstract

Throughout the last few years, permanent magnet synchronous motors have been proven suitable candidates for hybrid electric vehicles (HEVs). Among them, the outer rotor topology with surface mounted magnets and concentrated windings seems to be very promising and has not been extensively investigated in literature. In this study, an overall optimization and modelling procedure is proposed for the design and operational assessment of high-power density direct-drive in-wheel motors, targeted towards a light HEV application. The analytical model of an HEV's subsystems is then implemented for a more accurate evaluation of overall powertrain performance. Furthermore, a simple but effective cooling system configuration, which is taking into account the specific problem requirements, is also proposed.

Keywords: hybrid electric vehicle, high power density, in-wheel motor, optimization, permanent magnet motor, dynamic modelling, electrical machines design

1. Introduction

Recent environmental concerns due to global warming and air pollution motivated many countries around the world to legislate fuel economy and emission regulations for ground vehicles [1]. Furthermore, the necessity of developing alternative methods to generate energy for vehicles owing to depletion of conventional resources was greater than ever [2]. These features encouraged the introduction of fuel cell vehicles (FCVs), electric vehicles (EVs) and hybrid electric vehicles (HEVs) as suitable candidates for the replacement of the conventional internal combustion engine counterparts. Since the performance of EVs and FCVs is still far behind the requirements, HEVs are considered as the most reliable and preferred choice among similar technologies by manufacturers, governments and consumers [3, 4]. Comparing

INTECH
open science | open minds

© 2017 The Author(s). Licensee InTech. This chapter is distributed under the terms of the Creative Commons Attribution License (http://creativecommons.org/licenses/by/3.0), which permits unrestricted use, distribution, and reproduction in any medium, provided the original work is properly cited. (cc) BY to those technologies, HEVs are advantageous, as they exhibit high fuel economy, lower emissions, lower operating cost and noise, higher resale price, smaller engine size, longer operating life and longer driving range [5]. The world HEVs market has been rapidly growing and existing hybrid powertrains include passenger cars, small, medium and heavy trucks, buses, vehicles used in construction domain (e.g. forklifts, excavators), etc.

Despite HEV's high performance, their design and control strategies are not trivial. Multiple hybrid electric architectures have been developed and incorporated so far into commercially available vehicles in order to find acceptable design solutions with respect to various objectives and constraints [6]. Each configuration presents particular characteristics and the architecture selection depends on the application requirements and vehicle's type. For instance, series configuration is mainly used in heavy vehicles, whereas parallel-series one is preferable in small and medium automobiles, such as passenger cars and smaller buses, notwithstanding its more complex structure [7]. The specific topology combines the advantages of both series and parallel HEVs and has a greater potential in improving fuel economy and efficiency of the overall powertrain [8]. The HEV performance is even more enhanced when new design methodologies are implemented in order to find optimal configurations for power split devices, whereas at the same time, a single planetary gear is used [1].

However, the performance of an HEV is strictly dependent on the individual characteristics of its components (i.e. the internal combustion engine, the electrical motor and generator, the electronic equipment, the batteries, etc.). There is a strong "connection" among them and their collaboration interacts with the performance of the vehicle [9]. Several techniques, presented in [10], can be applied in order to achieve the optimal design and energy management of an HEV. According to [11], multi-objective optimization procedure and decision-making approach are necessary since there is a great amount of variables and goals to be taken into account. Moreover, among the most crucial decisions in the design of a HEV is the selection of the electric motor's type and its topology. A large amount of requirements such as (a) high power and torque density, (b) flux-weakening capability, (c) high efficiency over a wide range of speed, (d) high fault tolerance and overload capability, (e) high reliability and robustness, (f) low acoustic noise during operation and (g) low cost have to be met if a motor is to be considered as a suitable one for such an application [4].

Nowadays, various structures have been tested by HEV manufacturers and even more have been investigated in recent studies, e.g. [12]. Some of them, such as switched-reluctance motors (SRMs), despite their important advantages (high fault tolerance, simple construction, outstanding torque-speed characteristics and low cost) are currently not widely used for HEV applications. This is associated with the fact that they exhibit high acoustic noise, high torque ripple and low power factor [13]. Among the most popular topologies for this kind of traction system are induction and permanent magnet synchronous motors [14]. These two types are thoroughly examined and compared to each other [15] and their specific features are quite well known so far [16].

In order to meet the continuously increasing power density and efficiency requirements, PMSMs have become the dominant topology for light duty HEVs [14]. PMSMs with one or multiple layers of interior magnets fulfil the aforementioned characteristics and are commonly used in several commercial HEVs. Their typical output power varies from 30 to 70 kW for full hybrid passengers cars and can exceed 120 kW in the case of sport utility vehicles (SUVs). Recently, it has been found that surface-mounted permanent magnet synchronous motors (SPMSMs), especially when they are combined with concentrated windings instead of distributed ones, are also promising candidates for HEV propulsion [16]. They present high efficiency, satisfactory flux-weakening capability, low cogging torque and facile manufacturing procedure [17]. Honda Insight was one of the first commercial HEVs that incorporated this specific motor configuration. Since then, there has been increasing research interest for this topology.

That research effort though was carried out mainly for inner rotor topologies, in which the propulsion is provided by a single traction motor coupled with a gearbox and a differential. Thus, the perspective of mounting a motor with outer rotor to the wheel of a vehicle is very interesting and may present plenty of advantages. In this case, much lower flux density and respectively less magnet mass is required for the achievement of the same maximum torque. Copper as well as mechanical losses can be significantly lower than the corresponding ones of inner rotor topology. The manufacturing cost is lower, whereas at the same time, the total structure is lighter and can be constructed more easily [18]. Numerous in-wheel concepts for HEVs have been developed in the last years, mainly by Protean Electric and Mitsubishi.

The design procedure of direct-drive SPMSMs for an HEV presents increased complexity. There is a large number of variables and geometrical parameters that have to be estimated, while simultaneously numerous problem constraints have to be satisfied. The applied constraints refer to the maximum acceptable value of current density, the maximum value of dc-link voltage, the motor's volume and weight due to the limited available space, etc. Additionally, SPMSMs have to exhibit low-current harmonic content, non-saturable operation, low torque ripple and cogging torque. The determination of motor's thermal behaviour during different operating conditions and the implementation of the suitable cooling system are also of great importance. The adequate temperature alleviation can ensure the high driving performance, the motor's durability and the elimination of magnets demagnetization risk [19].

Based on the above, this chapter aims to investigate, optimize, compare and propose suitable high-power density in-wheel SPMSMs for a light HEV application. For this purpose, a design, optimization and modelling methodology for in-wheel motors is analytically presented in Section 2. According to this approach, the specifications of the derived topology are incorporated to an analytical HEV's model, which has been developed in Matlab/Simulink. In this way, the better approximation of the dynamic behaviour of the entire system is allowed. The performance estimation of each single subsystem and the calculation of parameters, such as the fuel consumption during different driving cycles, are also far more accurate. This methodology is compared to so far commonly used techniques, which are reviewed here too. Next, the proposed approach is applied to the case of two 15.3 kW in-wheel motors, which are going to be part of the driving system of a hybrid passenger car with series-parallel configuration. The derived results are given in Section 3 and relevant discussion is made regarding the motor and overall HEV system performance. Moreover, motors thermal behaviour is studied and a simple and effective cooling system for this kind of traction system is proposed. Finally, Section 4 summarizes and concludes the work.

2. High-power density direct-drive in-wheel motors

2.1. Requirements overview

The development of a direct-drive SPMSM, which will exhibit desirable performance, requires a large amount of problem variables, constants and constraints to be taken into account according to [20]. Moreover, meta-heuristic optimization techniques can be applied along with the classical design theory and the analytical equations. In this case, the multi-objective SPMSM optimization has to be modelled and performed carefully, especially when certain quantities are of primary concern [21]. The problem complexity is increased if an in-wheel PMSM is supposed to be incorporated into the powertrain of an HEV, whereas its operating point varies almost ceaselessly. Thus, the study of motor performance in the rated operating point or in the point of maximum provided torque, using finite element method (FEM) or fixed permeability method (FPM) has been proven inefficient enough [22]. Consequently, various design approaches and optimization methodologies have been revealed so far and each of them has its own advantages and disadvantages.

In classical HEV design process motor's efficiency map or torque-speed curve is a convenient way to represent drive system's performance. The determination of efficiency, torque and speed for different operating points permits the preliminary estimation of motor's characteristics in agreement with vehicle's attribute. Also, different topologies that are investigated as possible candidates for the same application can be easily compared to each other [23]. However, by using efficiency maps the motor is considered as a black box, which responds to certain inputs (voltage and current). These two variables are assumed to be optimal in order to achieve the highest efficiency at a specific torque and speed output. Furthermore, a map scaling factor model (MSFM), based again on the knowledge of an efficiency map, is generally used for the selection of motor's output power rating and specifications. The efficiency and torque are scaled using a linear dependency on the rated power. At the same time, few HEV's subsystems, such as the internal combustion engine, wheels, batteries and control scheme, can also be modelled constructing the appropriate equations and then a joint optimization of all the subsystems using dynamic programming can be performed [24].

Although the aforementioned procedure permits a better interaction between the electric motor/s and the other vehicle's subsystems, the approximation of the dynamic behaviour of the entire system is not satisfactory enough. It lacks accuracy concerning energy management estimation and fuel consumption calculation. Additionally, there is no association between motor's performance and its geometrical parameters. A compromise between FEA and MSFM method is introduced in [9], in which the detailed magnetic circuit model is incorporated in the optimization process. Starting from a preliminary topology, the final configuration can be derived when the user's requirements are met. The drawback of this approach is that only a restricted number of variables can be treated simultaneously. Thus, some geometrical parameters, such as motor's diameter and length, should be specified by the designer and this method should be applied only for the optimization of magnets and windings modulation. A fast magnetostatic FEA is proposed in [25] in order to address the specific problem. The derived results are now more precise and the computational time and complexity are significantly reduced. The final proposed PMSM configuration is developed studying motor's torque behaviour and minimizing stator flux linkage for the efficiency enhancement.

Another important requirement for the optimal HEV's operation is the minimization of motor's losses during different driving cycles or the overall profile of the HEV [26]. It is evident that design parameters that are optimized for one average assumed drive cycle are not necessarily optimal when an alternative use of the vehicle is carried out [27]. At least twelve characteristics points of representative driving cycles should be used for the analysis of motor's performance according to [28]. These points have to include acceleration, cruising and regenerative modes for more accurate fuel consumption calculation. A methodology based upon the overall driving cycle efficiency of the traction drive, which also takes into account the inverter losses, cooling system specifications and energy consumption of other subsystems, is presented in [29]. The implementation of the appropriate cooling system and the determination of its specifications are also of great importance as stated in [30].

2.2. Description of proposed methodology

The complex problem of the development of high-power density direct-drive SPMSMs for a light HEV can be solved by using a knowledge-based system (KBS), similar to that analytically described in [31]. The proposed architecture scheme (depicted in **Figure 1**) involves a number of knowledge sources (KS) and several layers that interact with each other, in order to ensure that the final solution is acceptable from technical, economical and manufacturing point of view. The first two layers (layer 0 and 1) incorporate the provided information about

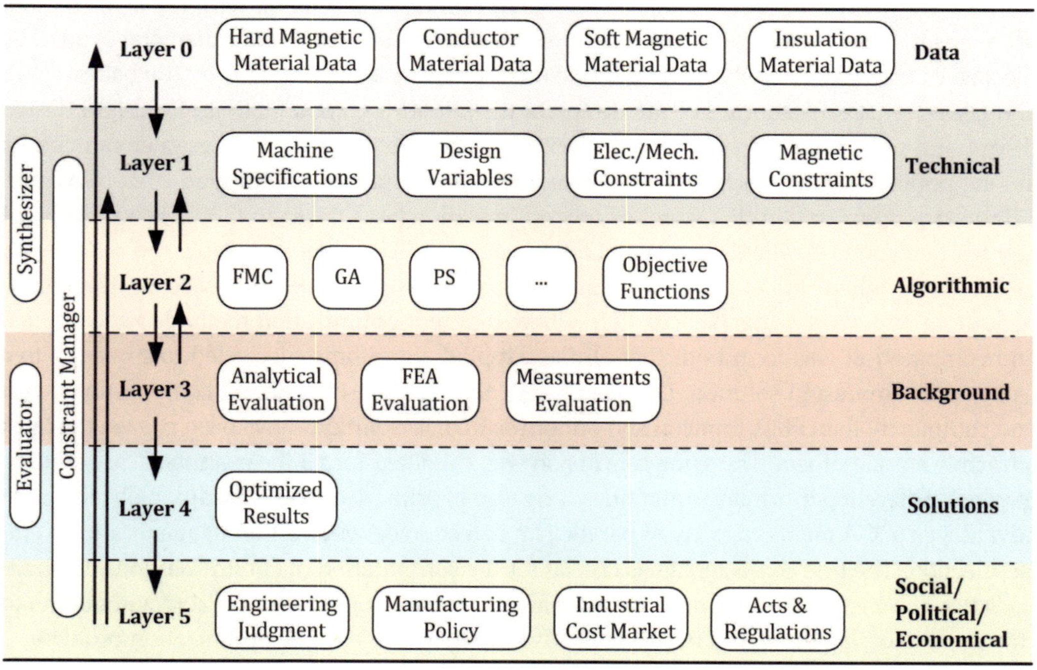

Figure 1. Structure of the developed knowledge-based system software.

the properties of high quality steels, soft magnetic conductors and insulation materials, while at the same time user's demands, machine's specifications, design variables and problem constraints are also determined. At the next level, the appropriate objective functions, taking into account the aforementioned, are constructed and an optimization method (e.g. genetic algorithm) is applied. At layer 3, an analytical evaluation of all the alternative derived solutions is conducted through FEA and post-processing analysis. Finally, the optimal motor configuration is selected (layer 4) and its application in HEV industry is thoroughly investigated.

The above approach was enhanced and finally an overall PMSM design and HEV performance assessment procedure is introduced in order to be a useful tool in the HEV design industrial process. This methodology is based upon the efficient design of the in-wheel motors and the determination of their average driving cycle efficiency. Furthermore, an analytical HEV's model, which has been developed in Matlab/Simulink, incorporates all the necessary subsystems of the vehicle. The internal combustion engine, the two identical SPMSMs coupled in the front wheels, the batteries pack, the dc-dc converter, the three-phase inverter, the power-split device and the control strategy are implemented in this model in order to permit a more realistic study of HEV's behaviour. For instance, the batteries model would make it possible to define the maximum provided voltage dynamically, while the state of their charge, the effect of their internal resistance, the effect of the prevailing temperature and working conditions can also be studied. Thus, a more appropriate selection of each single subsystem can be made resulting to an optimal energy management and performance.

The first step of the proposed methodology, which is presented in flowchart form in **Figure 2**, is the determination of motor's rated parameters, such as output power, speed and torque. These features are defined based on vehicle's speed and grade-ability along with the collaboration of in-wheel motors with the internal combustion engine. The outer motor diameter is fixed by the size of the wheel and the maximum dc-link voltage is also estimated by the battery pack and converter specifications. For the design of the SPMSMs a combination of classical design theory and meta-heuristic optimization techniques can be applied. The designer can choose among popular techniques based on swarm intelligence, such as genetic algorithm (GA), particle swarm optimization (PSO), ant colony optimization (ACO), etc. In [32], it is outlined that another new method called"Grey Wolf Optimizer" (GWO) exhibits acceptable and satisfactory performance when implemented in similar machine design problems. Based on the results of authors' previous works (i.e. [20, 21]), where different optimization methods were applied and compared, it was found out that all the adopted algorithms succeeded to converge to a (sub)-optimum design solution. Despite the fact that GA presents higher computational cost and complexity than PSO, fmincon and pattern search, its solutions have been proven the most attractive among others. The same conclusion was validated for all the examined case studies, in which different performance quantities were also of primary concern. Additionally, the main advantages of GA are its capacity of parallelism detection between different agents and its elitist selection. The first characteristic is crucial for the computation of Pareto solutions, whereas the latter one ensures that the best solutions are passed to the next iterative step without major changes. Following these, GA has been finally chosen for the specific optimization problem.

http://dx.doi.org/10.5772/intechopen.68455

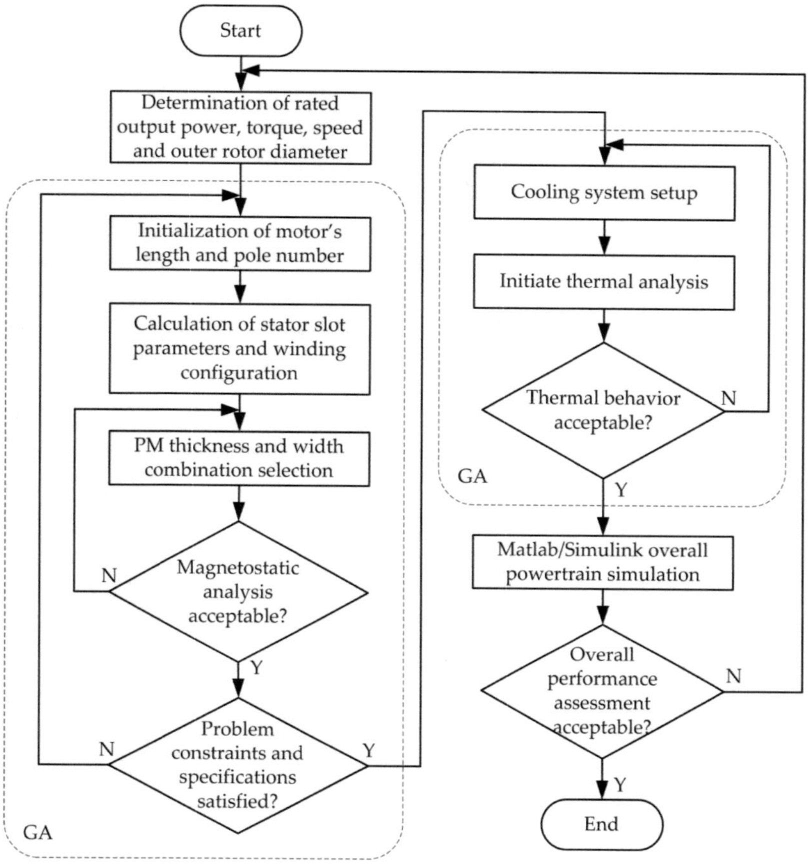

Figure 2. Proposed overall PMSM design and HEV performance assessment procedure.

Afterwards, the initialization of motor's length and poles number is following. The first parameter is specified by the size of the tire and the latter should be chosen carefully as it is of great importance for the overall motor performance [33]. The proper poles and slots combination can eliminate the presence of higher order harmonics in the air gap flux distribution, which results in lower iron losses and torque ripple. The motor must be capable of providing high torque when the vehicle accelerates while its losses should be as low as possible. Moreover, concentrated winding configuration is preferable for this traction system, since it presents shorter end windings and higher slot fill factor compared to the corresponding ones of distributed windings, and contributes to a smaller motor volume and lower copper losses, respectively. High power and torque density are very essential characteristics for such an application since there is restricted space inside the wheel. Furthermore, a high efficiency should be achieved over a wide speed range. Thus, the minimization of motor's volume simultaneously with the enhancement of efficiency will be of great concern during the construction of the objective functions. A weighted linear scalarization function is proposed—as a cost function—in order to"translate" the original multi-objective problem into a single-objective one which can be solved more easily. This function presents simplicity and thus the overall optimization complexity is reduced. Let the general form be $CF_j = \beta_i \cdot Q_i$, where β_i is a $1 \times i$ row matrix which contains the weight coefficients of the cost function and Q_i is an $i \times 1$ column matrix, which contains the values of any motor's quantities under optimization. Numerous cost functions can be produced in this way by altering the weights and/or quantities according to the problem specifications and user's requirements. Normally, a semi-exhaustive search has to be done first in order to explore weights search space for linear scalarization and, consequently, to identify efficient weight combinations. In the case examined here, we consider the motors' weight (M_{tot}) and efficiency (η) as equally important quantities for optimization, thus the above cost function is formulated as $CF = 0.5\,M_{tot} + 0.5\,(1 - \eta)$.

Next, the optimization procedure is applied for the determination of numerous variables, such as stator slot configurations, the number of turns per phase, the thickness and the width of permanent magnets, etc. At each step of the proposed approach, a large amount of constraints have to be met. Some of them are imposed in order to ensure the acceptable electromagnetic behaviour of the motor. For example, the motor's rotor yoke should be sufficient enough in order to ensure that no saturation will occur on this part of the machine. Likewise, the maximum acceptable flux density at other parts of the motor will also be set as problem constraints. The estimation of various electromechanical quantities using FEM analysis is indispensable in order to find out if any of these constraints is violated. If this happens, the adopted variables and geometrical parameters of the investigated topology have to be modified and the procedure returns to its initial step.

Another significant constraint is the maximum allowable value of the current density. For a totally enclosed in-wheel motor this value cannot exceed 10 A/mm^2 because there is no physical air circulation and temperature alleviation. Thus, the determination of this parameter and motor's thermal behaviour is essential in order to ensure high driving performance even under overload conditions, reduce the risk of magnets demagnetization and enhance the durability of insulation materials. Also, the implementation of a liquid cooling system for the motor is required. The research in recent literature revealed that the commonly used cooling system configurations are not suitable enough for this application. The oil-spray cooling method, which uses a radiator, is very energy consuming and increases the manufacturing complexity and the installation cost [34]. The implementation of ducting system and slot water jackets is difficult due to the limited space [35]. For the same reason, circumferential and axial water jackets are difficult to be applied, since the length of the motor is very short. Consequently, a more appropriate cooling system topology, which is effective enough despite the small cooling system surface, is developed and described thoroughly in the next Section. For each derived motor configuration, its thermal model and the thermal model of the proposed cooling system are constructed, the heat sources and the materials properties are specified, and the boundaries conditions and the temperature coefficients are determined. Finally, the temperature distribution and the overall performance of the cooling system are estimated. Its parameters are calculated by taking into account the optimal energy management of the HEV and the fact that the system's energy consumption must be kept as low as possible. The aim of the incorporation of motor's thermal analysis in the proposed methodology is to guarantee that motor designs

http://dx.doi.org/10.5772/intechopen.68455

which gather efficient performance and meet the specific problem requirements will not be excluded at this step of the design procedure due to over temperatures and high value of current density.

After applying all the eliminatory criteria, only optimal topologies are selected. Their geometrical parameters and specifications, such as stator phase resistance, inductance in d- and q-axis, flux linkage established by magnets, number of pole pairs, efficiency at rated power, source frequency, shaft inertia and damping coefficient, are then imported in Matlab/Simulink HEV model. This model, as mentioned before, involves all the necessary HEV subsystems and it will be used in order to assess the overall system performance. The final HEV configuration and motor topology will be chosen according to the optimal energy management and efficient collaboration of the subsystems. For this purpose, HEV performance can be estimated during one single or several different driving cycles. The designer should carefully choose the appropriate driving cycle, which fulfil his own requirements and the use of the vehicle. The urban driving cycle (ECE 15) and the New European Driving Cycle (NEDC) have been extensively employed by manufacturers for vehicle energy consumption and emission testing, as they represent the typical use of light duty vehicles in Europe.

Summarizing, the methodology proposed here seems to be very promising compared to other common practices, since it permits the detail implementation of motor's characteristics in HEV model and the interaction between its geometrical parameters with vehicle's performance. Additionally, the user can thoroughly compare to each other several candidate topologies before making his final choice, by examining aspects, such as the fuel consumption, the state of charge of the batteries, the compatibility of inverter's specifications with motor's requirements, etc. The large amount of constraints, the determination of motor's temperature distribution and electromechanical performance can ensure that the in-wheel motor will exhibit the desirable operation even under adverse working conditions. The relatively high simulation time that is required for running Matlab/Simulink model could be considered as the main disadvantage of the proposed here design procedure.

3. Case studies, results and discussion

In this Section the problem of the design and optimization of a light duty HEV's traction system is examined. The HEV under consideration incorporates the series-parallel configuration, using an internal combustion engine (ICE) and two SPMSMs for propulsion. The electric motors are implemented around each of the driving wheels to directly deliver power to them. Series-parallel architecture enables the engine and electric motors to provide power independently or in conjunction with one another. At lower vehicle's speeds the system operates more as series vehicle, whereas at high speeds, where the series drive train is less efficient, the engine takes over and energy loss is minimized. The engine is going to be able to produce 115 Nm torque at 4200 rpm, whereas its output power and its maximum speed will be 57 kW and 5000 rpm, respectively. The output power of each in-wheel motor will be equal to 15.3 kW and a torque of 170 Nm at 850 rpm will be provided. Moreover, the engine is going to drive a salient pole synchronous permanent magnet generator, which will either charge the batteries or provide power directly to the electric motors depending on vehicle's mode. A planetary gear is used in order to split power among the engine, the generator and the differential. The nominal voltage of the battery pack is 201.6 V comprising 168 nickel-metal hydride (NiMh) cells, and its nominal capacity is 6.5 Ah. The battery pack voltage is raised by a boost converter leading to a 400 V dc-link voltage. Finally, each in-wheel motor is fed through a three-phase inverter (**Figure 3**) and is individually controlled using vector control method. For the assessment of the overall system performance, a HEV model, which is available in Matlab/Simulink version R2016a (**Figure 4**) has been modified properly in order to meet the specific problem requirements. This model permits the study of vehicle's dynamic behaviour, as the aerodynamical and frictional phenomena are included. The vehicle's and its component specifications of the case study are presented in **Table 1**.

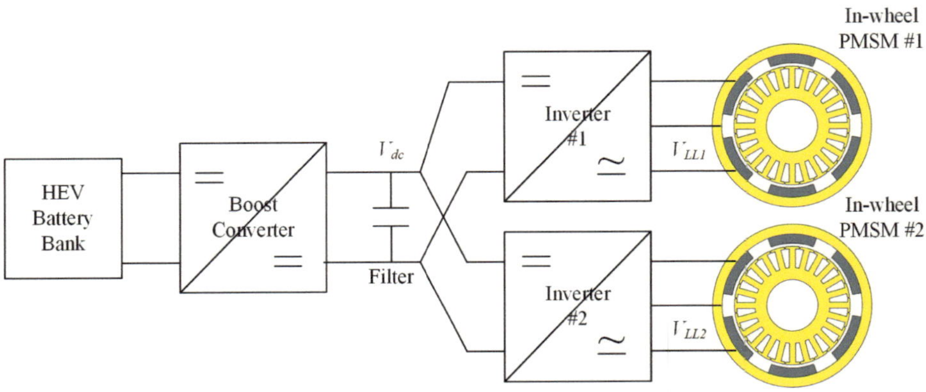

Figure 3. Typical in-wheel motors drive topology.

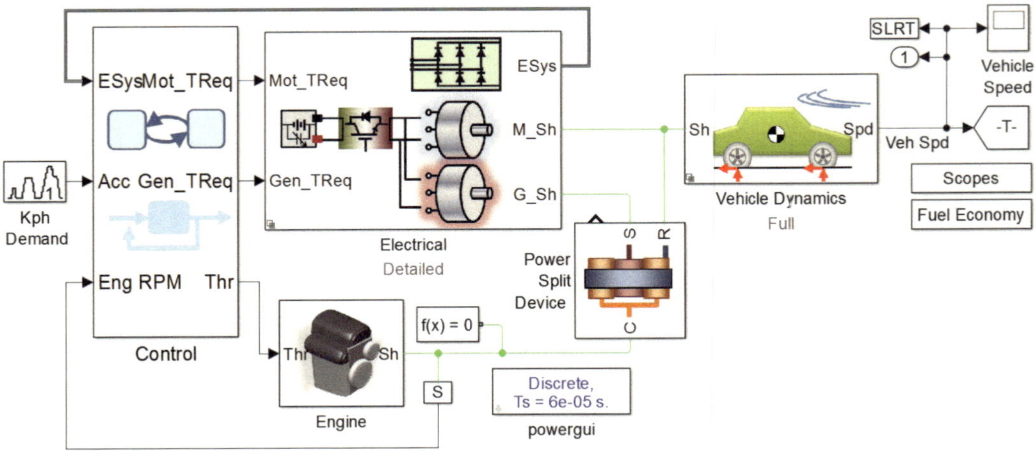

Figure 4. Matlab/Simulink HEV's model.

Component	Parameter	Value
Vehicle	Mass (kg)	1200
	Frontal area (m^2)	2.16
	Tire radius (m)	0.30
	Total wheel inertia (kg m^2)	0.10
	Aerodynamic drag coefficient	0.26
	Transmission inertia (kg m^2)	0.50
	Transmission friction coefficient	0.001
	Engine to wheel gear ratio	1.30
In-wheel motor (×2)	Rated power (kW)	15.3
	Rated speed (rpm)	850
	Rated torque (Nm)	170
	Rated power (kW)	57
Internal combustion engine	Maximum speed (rpm)	5000
	Torque (Nm) @ 4200 rpm	115

Table 1. HEVs under study main components specifications.

The design of a high-power density in-wheel motor is a complex optimization problem which will conclude to the most suitable candidates according to some criteria. There are several requirements that have to be met. Some of them are related to the motor's placement and physical constraints, such as its outer rotor radius and active length, whereas others are imposed by the motor's desired operation. The efficiency, for example, is of great importance considering the energy consumption. Efficiency higher than 90% will be an appropriate choice. Despite that, since the motor is mounted inside the wheel, as depicted in **Figure 5a**, its weight must be as low as possible in order to reduce unsprung mass and eliminate vibrations. Recently in-wheel motors with power-mass ratio of approximately 1 kW/kg have been implemented in commercially available HEVs. In this study, it will be investigated if this value can be exceeded. Thus, the objective function chosen for the case study is a compromise of motor's weight and power losses minimization. The desired SPMSMs characteristics are given in **Table 2**.

Furthermore, there are more than 15 design variables that have to be optimized simultaneously (under certain constraints) by the applied algorithm. Apart from the geometrical parameters that are presented in **Figure 5b**, variables such as the number of poles ($2p$), the number of slots per pole per phase (q) and the number of conductors per slot (n_c) are also involved. **Table 3** summarizes the upper and lower bounds of all these quantities that will be considered as problem constraints. At this point, it must be mentioned that for sake of space, the analytical equations that describe the electromechanical and magnetic behaviour of the specific machine are not given here. The reader can refer to [18, 36] for more details. Concerning the materials used for different motor's parts, a high quality silicon steel (M19-24G) has been selected both for stator and rotor, according to NEMA's instructions for super premium efficiency motors. Moreover, high energy NdFeB magnets have been chosen, as they have been proven efficient and reliable

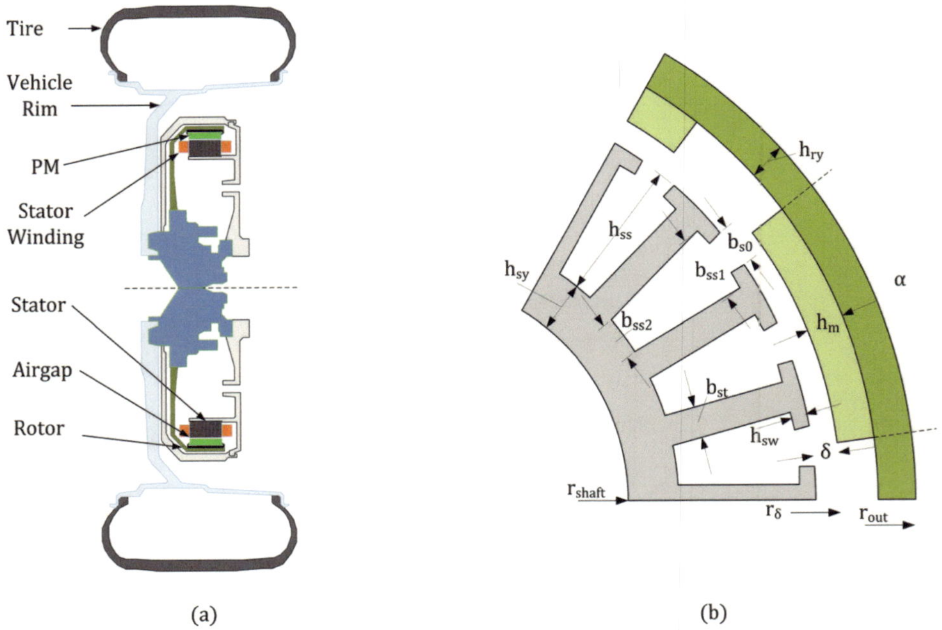

Figure 5. Representations of the design problem considered: (a) cross section of an in-wheel motor assembly and (b) detailed geometry of the SPMSM used here.

enough for this kind of application [37]. The values of materials properties will be regarded as constants during optimization problem and they are presented in **Table 4**.

Following the methodology proposed here, a set of optimization results are presented in **Table 5**, in which the design variables of four final solution topologies are given. Let us denote"Motor A" up to "Motor D" the derived in-wheel configurations. Their electromechanical performance has been validated through 2D and 3D FEM and considered acceptable and satisfactory enough. The obtained results are summarized in **Table 6**. Moreover, the overall HEV's system behaviour assessment has been conducted by examin-

Variable	Symbol	Value
Rated output power	P_{out}	15.3 kW
Rated output torque	T_{out}	170 Nm
Rated speed	n_s	850 rpm
Max. dc-link	V_{dc}	400 V
Inverter's module ratio	m_a	0.7
Axial length	L	30 mm
Outer rotor radius	r_{out}	216 mm

Table 2. In-wheel motor's desired characteristics.

Variable	Symbol	Value
Motor efficiency	η	≥90%
Magnets mass	M_{m}	≤30 mm
Motor total mass	M_{tot}	≤15 kg
Slot base width	b_{ss2}	$(0.15\,h_{ss} - 0.5\,h_{ss})$ mm
Slot opening width	b_{s0}	≥2.0 mm
Stator tooth width	b_{st}	≥2.5 mm
Magnet height	h_{m}	≤10 mm
Stator yoke height	h_{sy}	$\geq(h_{ss}/3)$ mm
Rotor yoke height	h_{ry}	≥9.0 mm
Air gap length	δ	(1–3) mm
Flux density in stator yoke	B_{sy}	≤1.8 T
Flux density in stator teeth	B_{st}	≤1.8 T
Flux density in rotor yoke	B_{ry}	≤1.8 T
Flux density in airgap	B_{δ}	≤1.0 T

Table 3. Optimization problem constraints.

ing vehicle's subsystems collaboration and calculating fuel consumption during four different driving cycles, which are depicted in **Figure 6**. Among the applied driving cycles are (a) European Driving Cycle ECE 15, (b) Extra Urban Driving Cycle (EUDC), (c) Supplemental Federal Test Procedure (SFTP-75) and (d) Japanese 10-15 mode driving cycle (JP 10-15). During ECE 15 cycle the vehicle covers a distance of 0.9941 km in 195 sec. The average and the maximum vehicle's speed are equal to 25.93 km/h and 50 km/h, respectively and its maximum acceleration is 1.042 m/sec^2. EUDC represents a more aggressive and high speed driving mode. The maximum developed and average speeds are equal to 120 km/h

Constant	Symbol	Value
Magnet density	ρ_{m}	7400 kg/m^3
Remanence flux density	B_{r}	1.24 T
Magnet relative permeability	μ_{r}	1.09
Copper density	ρ_{Cu}	8930 kg/m^3
Copper resistivity	r_{Cu}	1.72×10^{-8} Ωm
Steel density	ρ_{s}	7650 kg/m^3
Steel resistivity	r_{s}	5.1×10^{-8} Ωm
Steel relative permeability	μ_{s}	4000

Table 4. Optimization problem constants.

Variable	Symbol	Motor A	Motor B	Motor C	Motor D
Stator inner radius	r_{shaft}	159	159	162.47	155
Airgap radius	r_δ	199	199.12	200.84	199.12
Airgap length	δ	2	2	2.50	2.12
Number of poles	$2p$	28	34	48	60
Number of slots	Q_s	30	36	54	63
Slot opening width	b_{s0}	18.50	13.95	12.05	9
Slot top width	b_{ss1}	18.61	13.96	15	9.01
Slot base width	b_{ss2}	18.62	13.97	12.46	9.02
Stator tooth width	b_{st}	22.42	20.25	7.96	10.53
Slot height	h_{ss}	25	27	25.92	29
Stator tooth tip height	h_{sw}	1	1	1	1
Thickness of stator back	h_{sy}	13	11.12	9.95	13
Thickness of rotor back	h_{ry}	14.5	14.38	11.66	14.38
Magnet height	h_m	2.5	2.50	3.50	2.50
Pole arc/pole pitch ratio	α	0.79	0.78	0.60	0.90
Slot fill factor	s_f	0.6	0.6	0.60	0.60
Number of conductor per slot	n_c	34	26	16	14
Number of wires per conductor	n_w	1	1	1	1
Wire diameter	d_w	2.906	2.906	3.665	3.264
Number of layers	n_l	2	2	2	2
Winding factor	k_w	0.951	0.952	0.945	0.953

Table 5. Optimization design variables results and model comparison (all dimensions in mm).

and 69.36 km/h, respectively. SFTP-75 is commonly used for emission certification and fuel economy testing for light duty vehicles in United States. This cycle involves both driving in urban areas and high speed road. In this case study, only the second part of this cycle is incorporated. The duration of JP 10-15 is 660 sec and its average speed is 22.7 km/h. A fuel consumption lower than 5.0 l/100 km has been considered acceptable and each topology that

Quantity	Symbol	Motor A	Motor B	Motor C	Motor D
Motor efficiency (%)	η	94.41	94.31	94.97	95.41
Phase current (A)	I_{ph}	62.49	69.16	92.41	70.44
Current density (A/mm²)	J_c	9.42	9.68	8.76	8.41
Copper losses(W)	P_{cu}	839.81	864.75	722.06	604.61
Core losses (W)	P_{core}	66.93	69.16	87.75	132.44
Motor total mass (kg)	M_{tot}	14.82	14.36	12.49	14.77
Magnets mass (gr)	M_m	713.16	545.09	593.45	628.85
Cogging torque (mNm)	T_{cog}	41.07	50.96	31.05	23.15
Torque ripple (%)	T_{rip}	2.1	2.4	3.3	2.4
Torque angle (°)	T_{ang}	44.79	45.38	26.62	37.61
Nominal frequency (Hz)	f	198.3	240.8	340	425
Power density (kW/kg)	P_d	1.03	1.06	1.22	1.03

Table 6. Electromechanical quantities results (at rated condition).

did not meet the specific requirement has been excluded from the next step of the proposed methodology. The optimization procedure was terminated when for an investigated configuration the target was achieved for all the examined driving cycles. The relative results are presented in **Table 7**. From **Tables 5–7**, it is initially clear that the proposed approach succeeded in finding optimum and feasible design solutions satisfying all the existing constraints. Analytically the following can be observed:

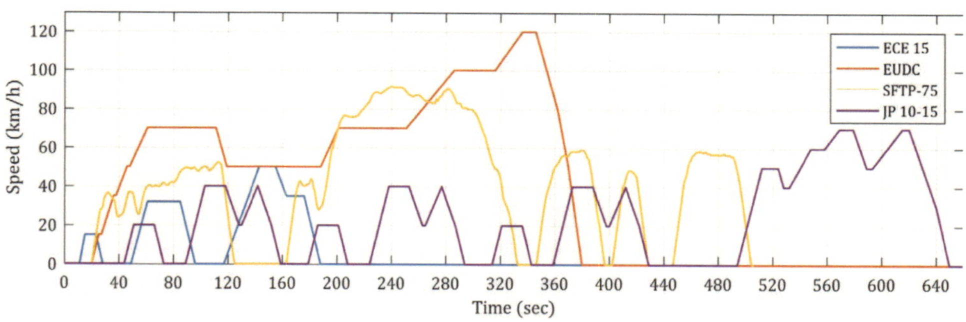

Figure 6. The four driving cycles used during the proposed optimization procedure.

(I) The optimization procedure provided solutions over the examined range of poles number and the final topologies are investigated and compared to each other from several aspects. The designer has the opportunity to evaluate the derived results from many points of view (i.e. technical, economical, etc.) and finally select the appropriate in-wheel SPMSM topology.

(II) The motors efficiency has been found high enough, as it varies from 94 to 95.5%. This feature, especially when is combined with the lowest possible current, is of great importance for HEV's energy management. Concerning this, Motor A seems to be a more suitable choice for the case study.

(III) All topologies exhibit high power to mass ratio over 1 kW/kg, since their mass range is from 12.5 to 14.8 kg. In the case of Motor C, the ratio is increased by 22%. Thus, if motor's total mass is the primary objective, this motor prevails. Despite their relatively low weight, all configurations present durability and do not suffer from mechanical stresses.

(IV) The volume of NdFeB magnets is small, which will lead to a reasonable motor's cost.

(V) The current density constraint has been fulfilled. However, concerning the short axial length of the machine (30 mm) and its placement into a totally enclosed environment the implementation of a cooling system, which has been also proposed and optimized here, is more than essential. More details about the cooling system's characteristics and its performance are going to be provided later in this chapter.

(VI) During the adopted design approach, a large amount of motor features were also determined, as they significantly affect its operation. Some of great importance estimated quantities are airgap flux density, torque and phase-back emf curve's shape, as well as their corresponding harmonics, cogging torque, torque angle and magnetic field distribution. For completeness purposes, these quantities are depicted in **Figures 7–11**, indicatively for Motor C and Motor D. As it can be seen from **Figure 7**, the values of flux density developed over the different parts of both configurations are found within acceptable limits. Despite the low volume and especially active length of the motor, non-saturable operation has been detected for all the finally proposed topologies. Moreover, the airgap flux density and the phase-back emf, as depicted in **Figures 8** and **9**, respectively, present low harmonic content. The proper selection of windings configurations along with the specification of permanent magnets parameters through the proposed approach contribute to this feature. The airgap flux is of great importance of the torque pulsation. The small amplitude of its third, fifth and seventh harmonic in both cases resulted in the low value of motors torque ripple. The torque ripple for Motor C was found equal to 3.3%, while the same parameter for Motor D was equal to 2.4%. The above can also be validated by the observation of **Figure 10**, in which the torque and its harmonic content is presented. A very low cogging torque and relatively torque angle is also achieved, as it can be seen from **Figure 11**. These parameters are essential for this kind of traction application, as their low value can ensure a high quality and safe driving performance.

(VII) The calculation of crucial HEV's parameters, such as fuel consumption, permits a better approximation of the optimal configuration. For example, Motor A seems to have a significant advantage compared to the other motors over all examined driving cycles, when the aspect of fuel consumption is examined. This may not be so clear if only the motor's rated performance characteristics were taken into account.

At this point, Motor C, which exhibit the higher nominal current compared to other topologies, is going to be used as a case study for the description of the applied cooling system. It must be outlined that a suitable cooling topology for this kind of motor should gather the following features: (a) be easily implemented in the restricted surface of the motor and (b) be close as much as possible to the part of the machine, which is the main heat source (i.e. stator copper windings). Taking these into account, the attachment of a cooling channel into the inner yoke circumference is proposed. This configuration, which looks like a ring, permits the circulation of a coolant through a pipe with rectangular cross section and the removal of the heat from the inner stator surface. A pump combined with a heat exchanger/compressor

Driving cycle	Quantity	Motor A	Motor B	Motor C	Motor D
	Fuel consumption (l/100 km)	3.83	4.19	4.76	4.27
ECE 15	Fuel consumption (km/l)	26.11	23.86	21.00	23.41
	Total fuel used (l)	0.0381	0.0417	0.0473	0.0424
	Fuel consumption (l/100 km)	2.68	2.81	3.42	2.99
EUDC	Fuel consumption (km/l)	37.31	35.59	29.24	33.44
	Total fuel used (l)	0.2064	0.2164	0.2634	0.2302
	Fuel consumption (l/100 km)	2.04	2.14	2.58	2.20
SFTP-75	Fuel consumption (km/l)	49.02	46.73	38.76	45.46
	Total fuel used (l)	0.1176	0.1234	0.1488	0.1274
	Fuel consumption (l/100 km)	3.39	3.50	3.94	3.46
JP 10-15	Fuel consumption (km/l)	29.49	28.54	25.41	28.94
	Total fuel used (l)	0.1180	0.1221	0.1371	0.1204

Table 7. Comparison of HEV fuel consumption using the designed in-wheel motors for different driving cycles.

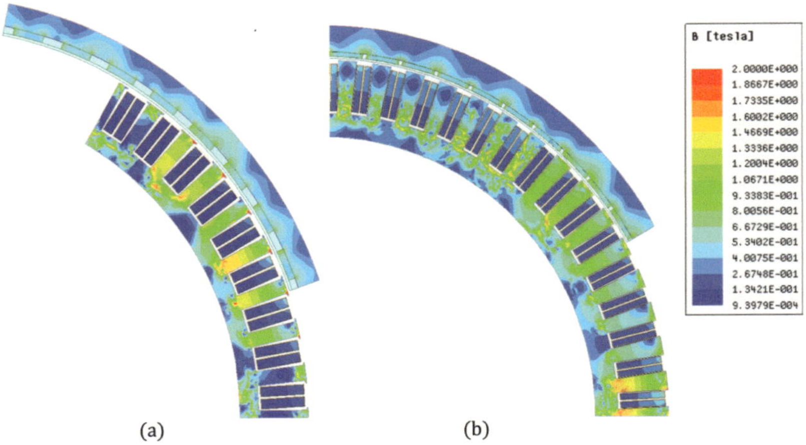

Figure 7. Flux density distribution in running conditions: (a) Motor C (48 poles/54 slots) and (b) Motor D (60 poles/63 slots).

for the alleviation of coolant's temperature will consist of the overall cooling system. The accurate position of the cooling channel is shown in **Figure 12**, in which the shell, the rim and the tire are also presented. All these will be parts of the developed thermal model in order to have a more accurate temperature determination. Moreover, during thermal analysis the temperature and the pressure inside the rim will be taken into account. Compared to other cooling system schemes, the proposed here topology enables a larger contact area between the stator and the coolant, simpler manufacturing and installation procedure and lower cost.

The design procedure of the proposed here cooling system requires the incorporation of an optimization method along with the conduction of motor's thermal analysis through FEM

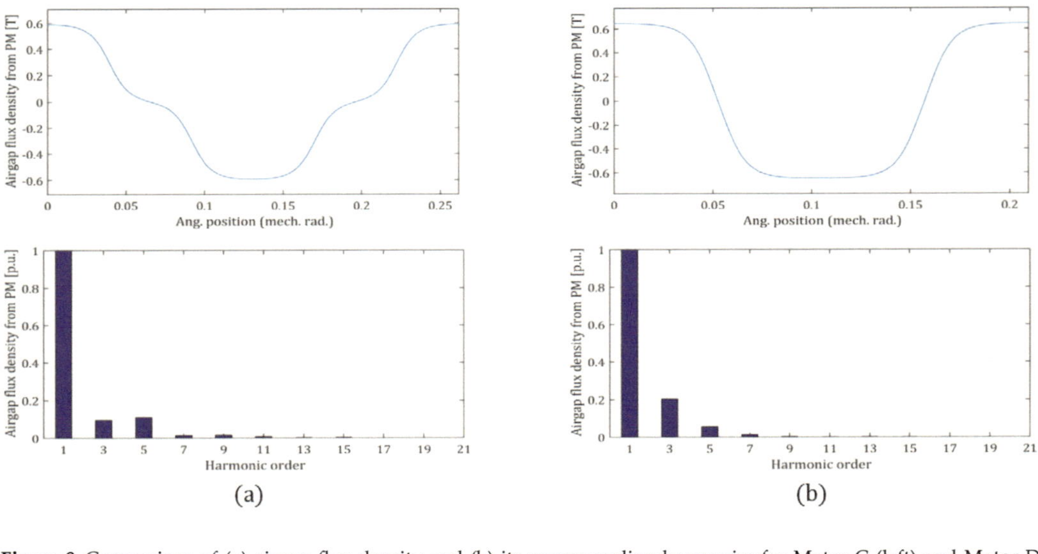

Figure 8. Comparison of (a) airgap flux density and (b) its corresponding harmonics for Motor C (left) and Motor D (right).

http://dx.doi.org/10.5772/intechopen.68455

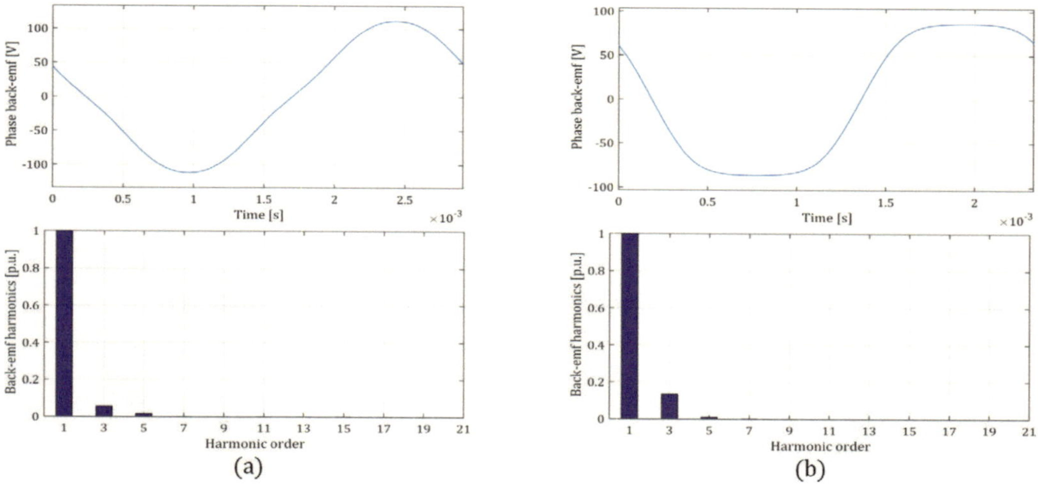

Figure 9. Comparison of (a) phase-back emf and (b) its corresponding harmonics for Motor C (left) and Motor D (right).

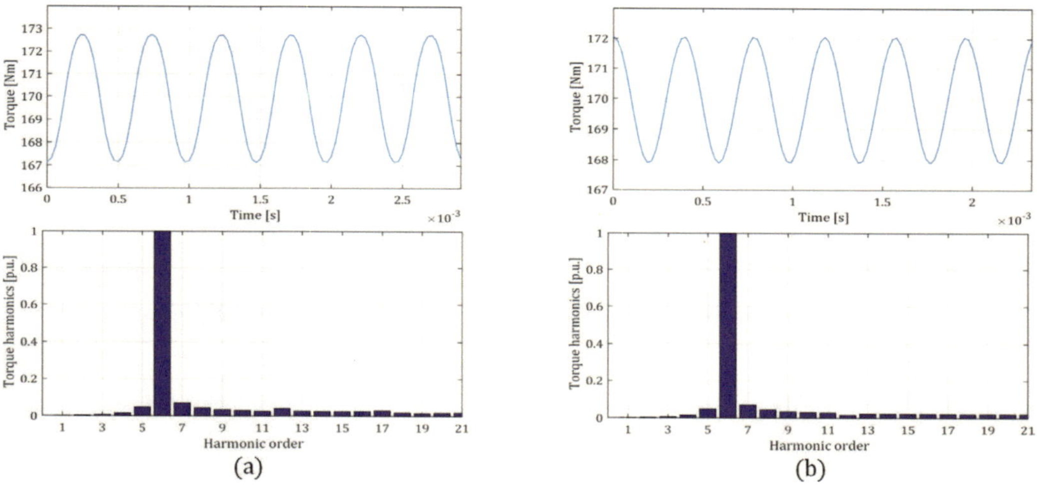

Figure 10. Comparison of (a) torque and (b) its corresponding harmonics for Motor C (left) and Motor D (right).

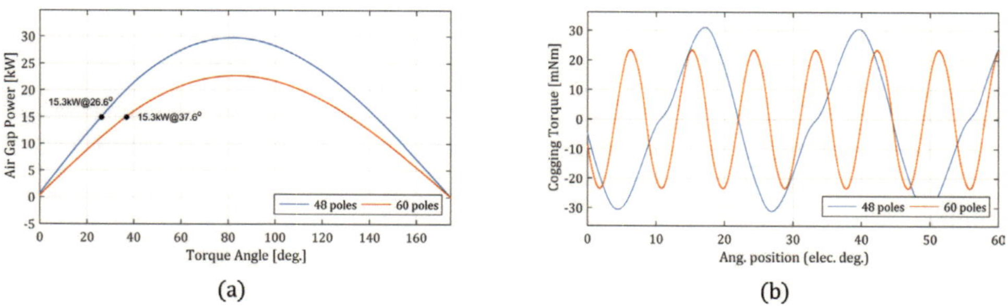

Figure 11. Comparison of electromechanical quantities for Motor C and D: (a) cogging torque and (b) airgap-developed power.

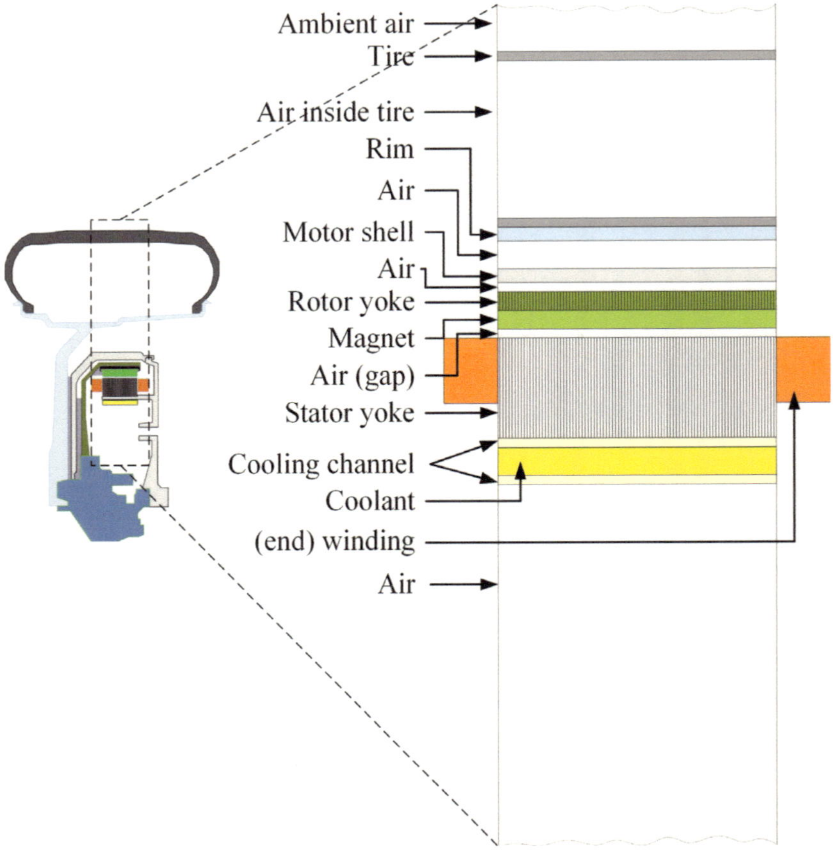

Figure 12. Cross section of the in-wheel motor topology, in which the accurate position of cooling channel is depicted.

under different operating conditions. The specifications of its parameters, such as the coolant's flow rate and coolant's inlet temperature are of great importance for system's efficiency and they will be calculated through the applied optimization algorithm. Since these parameters have essential effect on HEV's energy management, their values have to be carefully selected. For example, a high value for coolant's flow rate will conclude to increased energy consumption by the pump in order to circulate the liquid. Likewise, the heat exchanger should be capable of restoring coolant's temperature while its capacity will remain as low as possible. During this procedure, cooling channel's dimensions will be considered as variables. An aluminium alloy (6060-T6) with good mechanical properties has been selected for the channel, while ethylene-glycol mixed with water (50-50 volumetric proportion) has been chosen as coolant. The applied methodology involves the following the steps: (a) the determination of the thermal properties of all involved materials (including the shell, the tire and the insulation materials), (b) the specification of motor's heat resources and the ambient temperature, (c) the calculation of the boundary conditions in the air gap and other motor's parts and (d) the modification of the 2D thermal modelling according to the prevailing conditions.

Reader can refer to [38–40] in order to find more details about the classical theory governing the thermal analysis and the development of the motor's and cooling system's thermal model.

In **Figure 13**, the influence of coolant's inlet temperature and flow rate in temperature distribution over different in-wheel motor parts is presented. Based on these results and the aforementioned considerations, the inlet temperature of 30°C along with a flow rate of 4 l/min has been chosen as the optimal combination. Moreover, the channel's length, width and breadth have been specified to 30 mm, 10 mm and 1.5 mm, respectively. The derived requirements for heat exchanger, pump and pipe can be easily fulfilled by commercially available models. **Figure 14** shows the maximum observed temperatures of motor's parts for the same operating conditions without and with the application of the proposed cooling system. It can be easily observed that a significant temperature drop is achieved with the implementation of the cooling system at all the different loading conditions and the cooling system is considered efficient

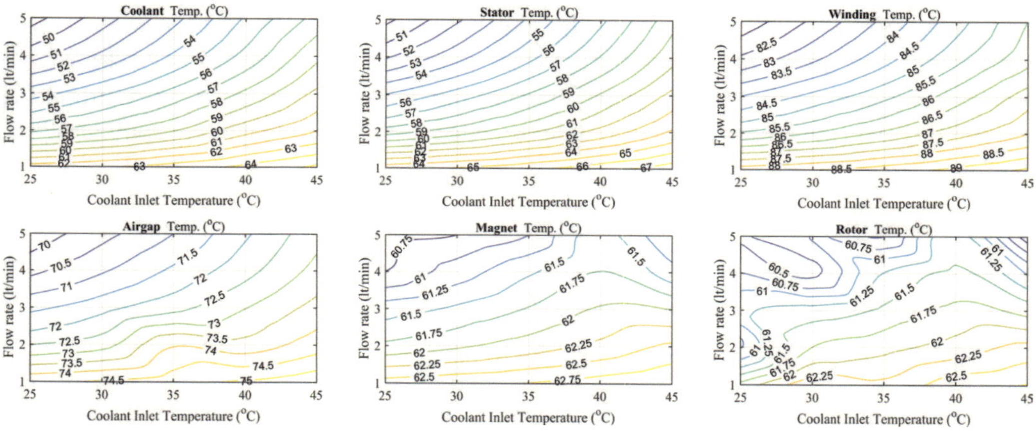

Figure 13. The influence of coolant's flow rate and inlet temperature on the temperature developed at different in-wheel motor parts.

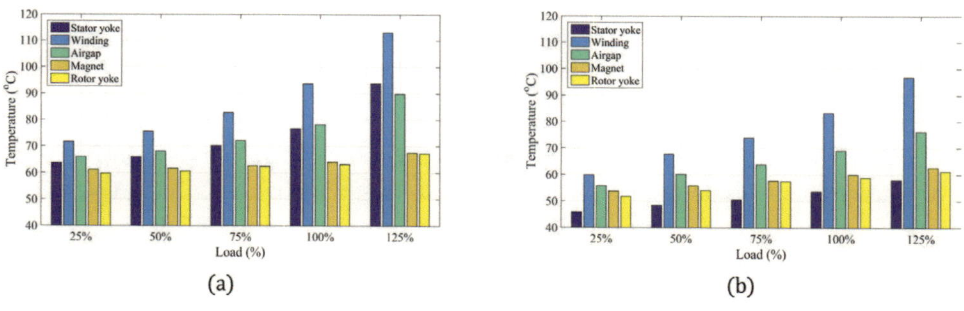

Figure 14. Comparison of in-wheel motor's temperature distribution: (a) without and (b) with the proposed here cooling system.

enough. The maximum heat removal is occurred in stator yoke and the copper windings. The maximum observed temperature in stator slots is far from the insulation materials limits. The temperature developed in air gap and especially near the magnets is relatively low and there is no risk of magnets demagnetization.

4. Conclusions

In this chapter, the perspective of direct-drive traction systems for HEVs, which lately concentrates on increasing interest among researchers and manufacturers but is not adequately investigated in the literature, is examined. A design and optimization methodology for the development of high-power density in-wheel motors and the corresponding beneficial assessment of the overall HEV's system performance is derived and discussed thoroughly. This approach is enhanced with the incorporation of a simple though efficient cooling system and the interaction of motor's geometrical parameters and performance with the vehicle's subsystems by using a dynamic HEV model. Through a case study, the particular problem requirements and constraints, the eliminatory criteria and the motor's topology selection strategy are illustrated and commented. Based on the overall results, the introduced methodology seems very promising and could be of great aid to designers in order to conclude to the optimal motor configuration.

Author details

Ioannis D. Chasiotis and Yannis L. Karnavas*

*Address all correspondence to: karnavas@ee.duth.gr

Department of Electrical & Computer Engineering, Electrical Machines Laboratory, Democritus University of Thrace, Xanthi, Hellas, Greece

References

[1] Kim H, Kim D. Comprehensive design methodology of input and output split hybrid electric vehicles. IEEE Transactions on Mechatronics. December 2016;**21**(6):2912-2913. DOI: 10.1109/TMECH.2016.2579646

[2] Kebriaeri M, Niasarm AH, Asaei B. Hybrid electric vehicles: An overview. In: Proceedings of International Conference on Connected Vehicles and Expo (ICCVE); 19-23 October 2015; IEEE, Shenzhen, China; 2015. pp. 299-305. DOI: 10.1109/ICCVE.2015.84

[3] Martinez CM, Hu X, Cao D, Velenis E, Gao B, Wellers M. Energy management in plug-in hybrid electric vehicles: recent progress and a connected vehicles perspective. IEEE Transactions on Vehicular Technology. Forthcoming. DOI: 10.1109/TVT.2016.2582721

[4] Chau KT, Chan CC, Liu C. Overview of permanent-magnet brushless drives for electric and hybrid. IEEE Transactions on Industrial Electronics. 2008;**55**(6):2246-2257. DOI: 10.1109/TIE.2008.918403

[5] Cosovic M, Smaka S. Design of initial topology of interior permanent magnet synchronous machine for hybrid electric vehicle. In: Proceedings of International Conference on Electric Machines & Drives (IEMDC); 10-13 May 2015; Coeur d'Alene, ID, IEEE, USA. 2015. pp. 1685-1664. DOI: 10.1109/IEMDC.2015.7409286

[6] Silvas E, Hofman T, Steinbuch M. Review of optimal design strategies for hybrid electric vehicles. In: Proceedings of 3rd IFAC Workshop on Engine and Powertrain Control, Simulation and Modeling, Elsevier; 23-25 October 2012; France. 2012. pp. 57-64. DOI: 10.3182/20121023-3-FR-4025.00054

[7] Chan CC, Bouscayrol A, Chen K. Electric, hybrid, and fuel-cell vehicles: Architectures and modeling. IEEE Transactions on Vehicular Technology. 2010;**59**:589-598. DOI: 10.1109/TVT.2009.2033605

[8] Cheng Y, Trigui R, Espanet C, Bouscayrol A, Cui S. Specifications and design of a pm electric variable transmission for Toyota Prius II. IEEE Transactions on Vehicular Technology. 2011;**60**(9):4106-4114. DOI: 10.1109/TVT.2011.2155106

[9] Vincent R, Emmanuel V, Lauric G, Laurent G. Optimal sizing of an electrical machine using a magnetic circuit model: Application to a hybrid electrical vehicle. IET Electrical Systems in Transportation. 2016;**1**(6):27-33. DOI: 10.1049/iet-est.2015.0008

[10] Yang H, Lin H, Zhu ZQ, Wang D, Fang S, Huang Y. A variable-flux hybrid-pm switched-flux memory machine for EV/HEV applications. IEEE Transactions on Industry Applications. 2016;**52**(3):2203-2214. DOI: 10.1109/TIA.2016.2524400

[11] Donateo T, Serrao L, Rizzoni G. A two-step optimisation method for the preliminary design of a hybrid electric vehicle. International Journal of Electric and Hybrid Vehicles (IJEHV). 2008;**1**(2):142-165. DOI: http://dx.doi.org/10.1504/IJEHV.2008.017831

[12] Chau KT, Li W. Overview of electric machines for electric and hybrid vehicles. International Journal of Vehicle Design. 2014;**64**(1):46-71. DOI: 10.1504/IJVD.2014.057775

[13] Williamson SS, Lukic SM, Emadi A. Comprehensive drive train efficiency analysis of hybrid electric and fuel cell vehicles based on motor-controller efficiency modeling. IEEE Transactions on Power Electronics. 2006;**21**(3):730-740. DOI: 10.1109/TPEL.2006.872388

[14] El-Refaie AM. Motors/generators for traction/propulsion applications: A review. In: Proceedings of International Conference on Electric Machines & Drives Conference (IEMDC); 15-18 May 2011; Niagara Falls, Ontario, Canada, IEEE; 2011. pp. 15-18. DOI: 10.1109/IEMDC.2011.5994647

[15] Pellegrino G, Vagati A, Boazzo B, Guglielmi P. Comparison of induction and PM synchronous motor drives for EV application including design examples. IEEE Transactions on Industry Applications. 2012;**48**(6):2322-2332. DOI: 10.1109/TIA.2012.2227092

[16] Wang J, Yuan X, Atallah K. Design optimization of a surface-mounted permanent-magnet motor with concentrated windings for electric vehicle applications. IEEE Transactions on Vehicular Technology. 2013;**62**(3):1053-1064. DOI: 10.1109/TVT.2012.2227867

[17] Pellegrino G, Vagati A, Guglielmi P, Boazzo B. Performance comparison between surface-mounted and interior PM motor drives for electric vehicle application. IEEE Transactions on Industrial Electronics. 2012;**59**(2):803-811. DOI: 10.1109/TIE.2011.2151825

[18] Karnavas YL, Chasiotis ID, Amoutzidis SK. Design considerations and analysis of in-wheel permanent magnet synchronous motors for electric vehicles applications using FEM. In: Proceedings of 17th International Symposium on Electromagnetics Fields in Mechatronics, Electrical and Electronic Engineering (ISEF); 10-12 September 2015; Valencia, Spain; 2015. Cd. Ref. no: 82

[19] Karnavas YL, Chasiotis ID, Peponakis EL. Cooling system design and thermal analysis of an electric vehicle's in-wheel PMSM. In: Proceedings 22nd International Conference on Electrical Machines (ICEM); 4-7 September 2016; IEEE, Lausanne, Switzerland; 2016. pp. 1439-1445. DOI: 10.1109/ICELMACH.2016.7732713

[20] Karnavas YL, Korkas CD. Optimization methods evaluation for the design of radial flux surface PMSM. In: Proceedings of 21st International Conference on Electrical Machines (ICEM); 2-5 September; IEEE, Berlin, Germany; 2014. pp. 1348-1355. DOI: 10.1109/ICELMACH.2014.6960357

[21] Karnavas YL, Chasiotis ID, Korkas CD, Amoutzidis SK. Modelling and multi-objective optimization analysis of a permanent magnet synchronous motor design. International Journal of Numerical Modelling: Electronic Networks, Devices and Fields. Forthcoming in 2017. DOI: 10.1002/jnm.2232

[22] Kwak SY, Kim JK, Jung HK. Characteristic analysis of multilayer-buried magnet synchronous motor using fixed permeability method. IEEE Transactions on Energy Conversion. 2005;**20**(3):549-555. DOI: 10.1109/TEC.2005.847973

[23] Dorrell DG, Knight AM, Popescu M, Evans L, Staton DA. Comparison of different motor design drives for hybrid electric vehicles. In: Proceedings of IEEE Energy Conversion Congress and Exposition (ECCE); 12-16 September 2010; IEEE, Atlanta, GA, USA. 2010. pp. 3352-3359 . DOI: 10.1109/ECCE.2010.5618318

[24] Sundstrom O, Guzzella L, Soltic P. Optimal hybridization in two parallel hybrid electric vehicles using dynamic programming. In: Proceedings of the 17th World Congress The International Federation of Automatic Control (IFAC); 6-11 July 2008; Elsevier, Seoul, Korea. 2008. pp. 4642-4647. DOI: http://dx.doi.org/10.3182/20080706-5-KR-1001.00781

[25] Ruuskanen V, Nerg J, Pyrhonen J, Ruotsalainen S, Kennel R. Drive cycle analysis of a permanent-magnet traction motor based on magnetostatic finite-element analysis. IEEE Transactions on Vehicular Technology. 2015;**64**(3):1249-1254. DOI: 10.1109/TVT.2014.2329014

[26] Lindh P, Tehrani MG, Lindh T, Montonen JH, Pyrhonen J, Sopanen JT, Niemela M, Alexandrova Y, Immonen P, Aarniovuori L, Polikarpova M. Multidisciplinary design of a permanent-magnet traction motor for a hybrid bus taking the load cycle into account. IEEE Transactions on Industrial Electronics. 2016;**63**(6):3397-3408. DOI: 10.1109/TIE.2016.2530044

[27] Schwarzer V, Ghorbani R. Drive cycle generation for design optimization of electric vehicles. IEEE Transactions on Vehicular Technology. 2013;**62**(1):89-97. DOI: 10.1109/TVT.2012.2219889

[28] Ruuskanen V, Nerg J, Parviainen A, Rilla M, Pyrhönen J. Design and drive-cycle based analysis of direct-driven permanent magnet synchronous machine for a small urban use electric vehicle. In: Proceedings of 16th Conference on Power Electronics and Applications; 26-28 Aug2014; IEEE, Lappeenranta, Finland; 2014. pp. 1-10. DOI: 10.1109/EPE.2014.6910915

[29] Tariq AR, Nino-Baron CE, Strangas EG. Design and analysis of PMSMs for HEVs based upon average driving cycle efficiency. In: Proceedings of IEEE International Conference on Electric Machines & Drives (IEMDC); 15-18 May 2011; Niagara Falls, IEEE, Ontario, Canada; 2011. pp. 218-223. DOI: 10.1109/IEMDC.2011.5994849

[30] Nguyen PH, Hoang E, Gabsi M. Performance synthesis of permanent-magnet synchronous machines during the driving cycle of a hybrid electric vehicle. IEEE Transactions on Vehicular Technology. 2011;**60**(5):1991-1998. DOI: 10.1109/TVT.2011.2118776

[31] Karnavas YL, Chasiotis ID. A simple knowledge base software architecture for industrial electrical machine design: Application to electric vehicle's in-wheel motor. In: Proceedings of 36th International Conference Information Systems Architecture and Technology (ISAT); 20-22 September 2015; Springer, Karpacz, Poland. 2015. pp. 111-122. Paper ID: 166189

[32] Karnavas YL, Chasiotis ID, Peponakis EL. Permanent magnet synchronous motor design using grey wolf optimizer algorithm. International Journal of Electrical and Computer Engineering. 2016;**6**(3):1353-1362. DOI: 10.11591/ijece.v6i3.9771

[33] Ahmada MZ, Sulaimana E, Harona ZA, Kosakab T. Impact of rotor pole number on the characteristics of outer-rotor hybrid excitation flux switching motor for in-wheel drive EV. In: Proceedings of 4th International Conference on Electrical Engineering and Informatics (ICEEI); 24-25 June 2013; Elsevier, Selangor, Malaysia; 2013. pp. 593-601. DOI: http://dx.doi.org/10.1016/j.protcy.2013.12.233

[34] Lim DH, Kim CS. Thermal performance of oil spray system for in-wheel motor in electric vehicles. Elsevier, International Journal of Applied Thermal Engineering. 2014;**63**(2):557-587. DOI: 10.1016/j.applthermaleng.2013.11.057

[35] Zhang B, Qu R, Wang J, Xu W, Fan X, Chen Y. Thermal model of totally enclosed water-cooled permanent-magnet synchronous machines for electric vehicle application. IEEE Transactions on Industry Applications. 2015;**51**(4):3020-3029. DOI: 10.1109/TIA.2015.2409260

[36] El-Refaie AM, Jahns TM, Novotny DW. Analysis of surface permanent magnet machines with fractional-slot concentrated windings. IEEE Transactions on Energy Conversion. 2006;**21**(1):34-43. DOI: 10.1109/TEC.2005.858094

[37] Kimiabeigi M, Widmer JD, Sheridan RS, Walton A, Harris R. Design of high performance traction motors using cheaper grade of materials. In: Proceedings of 8th IET International Conference of Power Electronics, Machines and Drives (PEMD); 19-21 April 2016; Glasgow, IET, UK; 2016. DOI: 10.1049/cp.2016.0287

[38] Lu Q, Zhang X, Chen Yi, Huang X, Ye Y, Zhy Q. Modeling and investigation of thermal characteristics of a water-cooled permanent magnet linear motor. IEEE Transactions on Industry Applications. 2015;**51**(3):2086-2096. DOI: 10.1109/TIA.2014.2365198

[39] Boglietti A, Cavagnino A, Staton DA. Evolution and modern approaches for thermal analysis of electrical machines. IEEE Transactions on Industrial Electronics. 2009;**56**(3): 871-882. DOI: 10.1109/TIE.2008.2011622

[40] Pyrhonen J, Jokinen T, Hrabovcova V. Design of Rotating Electrical Machines. John Wiley & Sons Ltd: Chichester, West Sussex, UK; 2013;612. DOI: 978-1-118-58157-5